Oscillations, waves and chaos in chemical kinetics

Stephen K. Scott

School of Chemistry,
University of Leeds

OXFORD NEW YORK TORONTO
OXFORD UNIVERSITY PRESS
1994

Oxford University Press, Walton Street, Oxford OX2 6DP

Oxford   New York   Toronto
Delhi   Bombay   Calcutta   Madras   Karachi
Kuala Lumpur   Singapore   Hong Kong   Tokyo
Nairobi   Dar es Salaam   Cape Town
Melbourne   Auckland   Madrid

and associated companies in
Berlin   Ibadan

Oxford is a trade mark of Oxford University Press

Published in the United States
by Oxford University Press Inc., New York

© Stephen K. Scott, 1994

All rights reserved. No part of this publication may be
reproduced, stored in a retrieval system, or transmitted, in any
form or by any means, without the prior permission in writing of Oxford
University Press. Within the UK, exceptions are allowed in respect of any
fair dealing for the purpose of research or private study, or criticism or
review, as permitted under the Copyright, Designs and Patents Act, 1988, or
in the case of reprographic reproduction in accordance with the terms of
licences issued by the Copyright Licensing Agency. Enquiries concerning
reproduction outside those terms and in other countries should be sent to
the Rights Department, Oxford University Press, at the address above.

This book is sold subject to the condition that it shall not,
by way of trade or otherwise, be lent, re-sold, hired out, or otherwise
circulated without the publisher's prior consent in any form of binding
or cover other than that in which it is published and without a similar
condition including this condition being imposed
on the subsequent purchaser.

A catalogue record for this book is available from the British Library

Library of Congress Cataloging in Publication Data
Scott, Stephen K.
Oscillations, waves and chaos in chemical kinetics/ Stephen K. Scott
(Oxford chemistry primers)

1. Chemical kinetics.   2. Oscillation chemical reactions.
3. Chaotic behavior in systems.   I. Title.   II. Series
QD502.S37   1994   541.3'94—dc20   93-48860
ISBN   0-19-855832-5 (Hbk)
ISBN   0-19-855844-9 (Pbk)

Typeset by the author and AMA Graphics Ltd., Preston, Lancs
Printed at Alden Press Limited,
Oxford and Northampton, Great Britain

Series Editor's Foreword

Oxford Chemistry Primers are designed to provide clear and concise introductions to a wide range of topics that may be encountered by chemistry students as they progress from the freshman stage through to graduation. The Physical Chemistry series will contain books easily recognized as relating to established fundamental core material that all chemists will need to know, as well as books reflecting new directions and research trends in the subject, thereby anticipating (and perhaps encouraging) the evolution of modern undergraduate courses.

In this second Physical Chemistry Primer, Stephen Scott has produced an exhilarating, easy-to-read introduction to Oscillations, Waves and Chaos in Chemical Kinetics. This Primer will interest all students (and their mentors) who wish to appreciate the importance of nonlinear kinetics in a wide diversity of chemical phenomena.

Richard G. Compton
Physical Chemistry Laboratory, University of Oxford

Preface

The subject of 'nonlinear dynamics' and the phenomenon of chaos, in particular, has become of great interest across of wide range of scientific disciplines in recent years. Chemical kineticists have been very much involved in these developments. Chemical reactions are exemplary nonlinear systems and have provided some of the most strikingly visual experimental evidence and some of the 'neatest' models for theoretical studies. In general, however, much of the chemical community still regards such phenomena and those who study them as mavericks. This text sets out to provide a simple introduction to the excitement and relevance of nonlinear chemical kinetics to the chemical world in general. Many examples turn out to involve old friends such as chain branching, autocatalysis and the Arrhenius temperature dependence of rate constants. These forms of 'chemical feedback' give rise to a hierarchy of exotic behaviour from clock reactions to the most complex aperiodicities. Links to other subject areas such as combustion, biochemistry and physiology are also demonstrated.

The text was written during a period of study leave in the Department of Chemistry, at West Virginia University. It is a pleasure to thank that Department and, in particular, my host Professor Ken Showalter, for their hospitality, the School of Chemistry at University of Leeds for allowing me to take leave, and the Council for the International Exchange of Scholars (Fulbright Commission) for financial support for travel. It is also a pleasure to thank the staff at Oxford University Press for their usual professional support, and Hilary and Lucy for personal support.

Contents

1 Introduction to basic concepts — 1
2 Clock and fronts — 14
3 Resetting the clock: oscillations — 26
4 Targets, spirals, and scrolls — 41
5 Bistability: steady states in flow reactors — 53
6 Oscillatory reactions in flow systems — 67
   Further reading — 91
   Index — 92

1 Introduction to basic concepts

The phenomena of this book, such as oscillations, ignitions, and chemical patterns, have become of great interest across the world chemistry community in recent years. There is, however, a long history that has unfolded in this area paving the way for the explosion of activity of the last two decades. Table 1.1 indicates something of a personal choice of 'significant events' and begins with observations accredited to Robert Boyle in the seventeenth century. Boyle noted a periodic 'flaring-up' of phosphorus in a loosely stoppered flask arising from the interaction of chemical kinetics and diffusion. The reaction between phosphorus and oxygen is a branched chain process that leads to an ignition. The ignition consumes the oxygen available in the flask and so the reaction ceases. As more oxygen diffuses into the flask, the reaction does not immediately recommence. Instead, the oxygen concentration must reach a 'critical' value before the chain branching leads to another ignition process. Other isolated observations were reported up to the beginning of the present century in a range of chemical and physical systems. The Landolt clock reaction (1886) has become a staple example of interesting kinetics and will be one of the systems discussed in this book.

A.M. Zhabotinsky (1991). A history of chemical oscillations and waves. *Chaos*, **1**, 379–86.

A 'Dark Ages' period between approximately 1900 and 1960 saw theoretical chemists hold in disdain any claims that chemical reactions could oscillate. There was a firm conviction that the second law of thermodynamics would just not allow that sort of abnormality—at least in homogeneous systems. Perhaps the greatest victim of that completely incorrect prejudice was Boris Belousov, whose attempts at publication of the observations of oscillations in the reaction that now bears his name jointly with Anatol Zhabotinsky were totally frustrated by those who 'knew better'. The new dawn of the modern period arose with the joint emergence of irrefutable experimental evidence of Belousov's oscillations by Zhabotinsky and the breaking of the theoretical chains by Ilya Progogine and his group in Brussels. (It might also be noted that chemical engineers and combustion scientists who did not know that chemists 'knew' that oscillations were impossible had been happily making considerable progress in this area since 1945.)

This story is told by A.T. Winfree (1984). The prehistory of the Belousov–Zhabotinsky reaction. *J. Chem. Educ.*, **61**, 661–3.

In the last two decades, a great deal of mastery has been achieved that allows oscillatory reactions almost to be designed at will. This book introduces many of the phenomena and their simplest chemical interpretation. The subject area is that of 'applied chemical kinetics', in some cases coupled to fluid flow or molecular diffusion. The remainder of this chapter revises some standard chemical kinetics and introduces the more specialist concepts of *nonlinearity* and *feedback*. Specific examples of these are

Introduction to basic concepts

Table 1.1 Some important dates in nonlinear chemical kinetics

Period	Experiment	Theory
late 1600s	Boyle: oscillatory P ignition	
early 1800s	Fechner: oscillatory dissolution Davy: cool flames	
1830s	Munck: oscillatory P ignition	
1886–1905	Landolt, Dushman, Roebuck: iodate clock	
1910s	Morgan: gas evolution oscillators Heathcote: iron nerve Bredig: beating Hg heart	Luther: waves Liljenroth: nonisothermal CSTR Lotka: biological model for oscillations
1920–30s	Bray: iodate–H_2O_2 oscillations	Fisher; Kolmogorov: waves
1940s	Newitt and Thornes: cool flames in propane oxidation	Zel'dovich: isolas and mushrooms Denbigh: autocatalytic reactions Sal'nikov: model for thermokinetic oscillations
early 1950s	Ashmore and Norrish: CO oscillations (batch)	Turing: pattern formation
late 1950s	Belousov: BZ reaction (batch) oscillations	Aris and Amundson: stability analysis
early 1960s	Zhabotinsky: BZ (batch and waves) Ghosh and Chance: glycolytic oscillations	Prigogine and Nicolis: Brusselator model Lorenz: chaos
late 1960s	Linnett et al.: CO oscillations (batch)	Gray and Yang: local stability analysis for hydrocarbons Selkov model for glycolytic oscillations
1972–4	Field Körös, Noyes: BZ mechanism Winfree: BZ spiral waves	Oregonator model
1973	Briggs–Rauscher oscillating reaction	
late 1970s	Epstein, De Kepper, Orban, Kustin: design of oscillators Schmitz and Hudson: BZ chaos and mixed-mode oscillations Degn and Olsen: chaos in enzyme systems Leeds groups: CO, H_2 and hydrocarbons (CSTR)	Boissonade and De Kepper: cross-shaped diagrams (CSTRs) May: chaos in maps Rossler: chaos Feigenbaum: chaos
early 1980s	Texas BZ chaos Showalter and Ganapathi: isolas and mushrooms Epstein and Orban: chlorite–thiosulfate chaos Stanford, Dortmund and Bordeaux groups: imaging of spirals Hudson: electrodissolution oscillations	Tyson and Keener: BZ spirals and scrolls Leeds groups: isolas and mushrooms in autocatalytic models Balakotaiah & Luss: singularity theory
late 1980s	De Kepper, Boissonade, Ouyang and Swinney; Epstein and Lengyel: unstirred flow reactors and pattern formation	Field and Gyorgyi: models of BZ chaos
1990–	Leeds group: chaos in CO + O_2 reaction	Showalter group: controlling chaos

discussed in terms of the three chemical reactions that will be used throughout this book: the Belousov–Zhabotinsky reaction, the Landolt clock reaction and the gas-phase oxidation of hydrogen.

1.1 Mechanisms and rate laws: law of mass action

The quantitative interpretation of experimental results in chemical kinetics is frequently based on the construction and subsequent analysis of *mechanisms* or *models*. Mechanisms typically aim for a semiquantitative match to a particular reaction; models are often deliberately simpler and attempt to catch the main qualitative features of a broader class of reactions.

Elementary steps

If we propose a sequence of individual reactions which are likely to be significant in carrying the reaction from the original reactant to the final products, the corresponding *reaction rate equations* can then be written out based on the *law of mass action*. The reaction rate equations specify the rates at which the concentrations of the various chemical species change with time, and how these rates depend upon those concentrations. The idea is to break the overall reaction up into a number of component reactions. Typical of such a component is an *elementary step*. An example of an elementary step is the following, between a hydrogen atom and an oxygen molecule:

$$H + O_2 \rightarrow OH + O. \tag{1.1}$$

The rate of an elementary step is given simply in terms of the concentrations of the participating reactants (those on the left-hand side of the reaction step). For the above example:

$$\text{Rate} = k[H][O_2] \tag{1.2}$$

Elementary steps are typically, but not exclusively, *bimolecular*—involving two species. The reaction is envisaged as occurring in a single collision between the reactants. Bimolecular reactions then give rise to a quadratic dependence of rate on concentration, i.e., to an overall *second-order* process. Equation (1.2) also involves a coefficient k. This is the *reaction rate coefficient* or *reaction rate constant*. Typically, rate constants are independent of concentration but they may be quite sensitive functions of the temperature. This temperature dependence can frequently be expressed in the form

$$k(T) = A \exp(-E/RT) \tag{1.3}$$

where T is the thermodynamic (absolute) temperature and R is the universal gas constant ($R = 8.314$ J K^{-1} mol^{-1}). The parameter A is termed the *pre-exponential factor* and E is known as the *activation energy*. Many chemical reactions are exothermic or endothermic. The consequent evolution or removal of energy may cause local temperature rises in the reacting mixture

4 Introduction to basic concepts

and the Arrhenius temperature dependence of k may then have important consequences.

Overall rate equations

The overall rate equations for a given mechanism are derived by combining the rates of all the individual elementary steps. The hydrogen–oxygen reaction is an example that will be used in various places in this book. An acceptable mechanism for this system over a limited (but very interesting) range of pressure and temperature involves the following elementary steps:

(0) $H_2 + O_2 \rightarrow 2OH$ Rate = $k_0[H_2][O_2]$

(1) $OH + H_2 \rightarrow H_2O + H$ Rate = $k_1[OH][H_2]$

(2) $H + O_2 \rightarrow OH + O$ Rate = $k_2[H][O_2]$

(3) $O + H_2 \rightarrow OH + H$ Rate = $k_3[O][H_2]$

(4) $H \rightarrow \tfrac{1}{2}H_2$ Rate = $k_4[H]$

In each case, the rate of the individual step has been derived from the law of mass action. The rate is defined as the rate at which the reactant species disappears. In the last step, this means that a noninteger *stoichiometric factor* arises.

The reaction rate equations can now be constructed. As an example, we can consider the rate of change of the hydroxyl radical concentration, d[OH]/dt. Two OH radicals are produced in the first step and one is produced in each of steps (2) and (3), whilst an OH radical is removed in step (1). The rate of equation is thus

$$\frac{d[OH]}{dt} = 2k_0[H_2][O_2] - k_1[OH][H_2] + k_2[H][O_2] + k_3[O][H_2] \quad (1.4)$$

The full set of governing rate equations for the six different chemical species in this mechanism is shown in Table 1.2. A number of general features can be noted from these equations. Any removal term (i.e., any term preceded by a minus sign) involves the concentration of the species being removed. Chemically, this is almost self-evident: a species must be directly involved in any step that consumes it. An important consequence of this is that chemical concentrations will not become negative. If the concentration of any given species becomes zero, then all the removal terms also vanish, so the rate of change of that concentration must then be either zero or positive.

A second feature typical of most elementary steps is that the production terms in any rate equation do not involve the concentration of the species being produced. This is not always the case, and becomes more common when we move to empirical rate equations in later sections.

Although there are six rate equations listed in Table 1.2, the concentrations of the six chemical species cannot all vary independently. There are two different types of chemical building blocks, the H atom and the O atom

Table 1.2 Governing rate equations for the $H_2 + O_2$ mechanism (0)–(4)

$$\frac{d[H_2]}{dt} = -k_0[H_2][O_2] - k_1[OH][H_2] - k_3[O][H_2] + \tfrac{1}{2}k_4[H]$$

$$\frac{d[O_2]}{dt} = -k_0[H_2][O_2] - k_2[H][O_2]$$

$$\frac{d[H]}{dt} = k_1[OH][H_2] - k_2[H][O_2] + k_3[O][H_2] - k_4[H]$$

$$\frac{d[OH]}{dt} = 2k_0[H_2][O_2] - k_1[OH][H_2] + k_2[H][O_2] + k_3[O][H_2]$$

$$\frac{d[O]}{dt} = k_2[H][O_2] - k_3[O][H_2]$$

$$\frac{d[H_2O]}{dt} = k_1[OH][H_2]$$

from which all the species are formed. As no reaction converts H to O, or vice versa, the total number of each of these atoms must be conserved. Thus, here there are two additional *conservation conditions*: if the initial composition is that of pure H_2 and O_2 at concentrations $[H_2]_0$ and $[O_2]_0$ respectively, then at all times we have

$$2[H_2] + [H] + [OH] + 2[H_2O] = 2[H_2]_0 \quad (1.5)$$

and

$$2[O_2] + [O] + [OH] + [H_2O] = 2[O_2]_0 \quad (1.6)$$

Thus if [H], [O], [H_2], and [OH] are known, [H_2O] and [O_2] are constrained to have the appropriate values to satisfy Eqns (1.5) and (1.6). With six rate equations and two conservation conditions, the system has four *degrees of freedom* or four *independent concentrations*.

Nonelementary processes

In many real situations, a kinetic mechanism involving only true elementary steps is not attainable—or always needed. Often a number of elementary steps are so intimately coupled that it is only possible to observe their overall effect: we refer then to an *overall process*. In these situations, experimentally-determined *empirical rate laws* are particularly useful. The Belousov–Zhabotinsky (BZ) reaction will feature prominently in later chapters. As a part of the mechanism for that system, the following processes are important

(5) $\quad BrO_3^- + Br^- + 2H^+ \to HBrO_2 + HOBr$

(6) $\quad HBrO_2 + Br^- + H^+ \rightarrow 2HOBr$

(7) $\quad HOBr + Br^- + H^+ \rightarrow Br_2 + H_2O$

None of these is necessarily believed to be an elementary step even though they have each been written in the same stoichiometric form as the elementary steps seen above in the $H_2 + O_2$ mechanism. Here, the form of reactions (5)–(7) is partly to identify the species that disappear in each process (the 'reactants') and those that are formed (the 'products'). In some cases, this form is also partly intended to convey information about the empirical rate laws which have been determined for each particular process. Thus, for process (5) the rate of conversion of bromate or bromide ions into the species $HBrO_2$ and $HOBr$ is found to be first order in $[BrO_3^-]$ and in $[Br^-]$ and second order in the concentration of H^+; thus we can write for process (5)

$$-d[BrO_3^-]/dt = -d[Br^-]/dt = -\tfrac{1}{2} d[H^+]/dt = +d[HBrO_2]/dt$$
$$= +d[HOBr]/dt = k_5[BrO_3^-][Br^-][H^+]^2 \qquad (1.7)$$

There is no suggestion, however, that this conversion is brought about by a single reactive collision between the four species on the left-hand side of reaction (5). In general, it is wise to quote the empirical rate law with the reaction step, as the law of mass action form does not always apply. Once all the important overall processes and their empirical rate laws have been identified, these are again combined to produce the full rate equations for the total rates of change of each concentration.

1.2 Nonlinearity and feedback: twin impostors

The buzzwords 'nonlinearity' and 'feedback' are the key features of the chemical kinetics underlying all the exotic phenomena to be described in this book. In this section, we will see that both are really rather common and 'normal' features of chemical reactions.

Nonlinearity

As mentioned in the previous section, the simplest type of chemical reaction is the elementary step, which we might represent with the general form

$$A + BC \rightarrow AB + C \qquad (1.8)$$

with

$$-d[A]/dt = -d[BC]/dt = d[AB]/dt = d[C]/dt = k[A][BC] \qquad (1.9)$$

In many kinetic studies aimed at determining the reaction rate constant k, the concentration of one of the two reactions, BC say, will be arranged to be in great excess over that of the other. The concentration of the species in excess can then be treated as a constant, giving a pseudo-first-order reaction. This is the only example of linear system in chemical kinetics. All non-first-order processes are nonlinear and thus *nonlinearity is the rule*, rather than the exception.

The meaning of nonlinearity is particularly easily visualized if we plot the variation of the reaction rate with the *extent of reaction* ξ. The latter is related to the instantaneous concentration of the reactant such that ξ varies between 0 (at the beginning of the reaction) and 1 (corresponding to complete conversion to products). Thus for the above equation, we may take $\xi = ([A]_0 - [A])/[A]_0$ where $[A]_0$ is the initial concentration. The reaction rate can now also be explicitly defined as $d\xi/dt = (1/[A]_0)(d[A]/dt)$. Figure 1.1 shows $d\xi/dt$ as a function of ξ for a various possible rate laws. For first-order reaction, $d\xi/dt = k(1 - \xi)$. This gives a straight line, i.e., a linear relationship. If instead of choosing BC to be in excess, the two reactants begin with equal concentrations, we would find simple second-order kinetics, $d\xi/dt = k(1 - \xi)^2$, giving a parabolic and hence *nonlinear* plot. Also shown is a curve corresponding to half-order kinetics, such as may arise from an empirical rate law for a nonelementary process: again, the dependence is not linear.

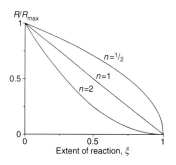

Fig. 1.1 The dependence of reaction rate $R = (-d\xi/dt)$ on extent of conversion ξ for half-order ($n = \frac{1}{2}$), first-order ($n = 1$) and second-order ($n = 2$) kinetics. The rate is normalized by the maximum rate R_{max} which, for these deceleratory systems, corresponds to the initial rate with $\xi = 0$.

Feedback

Nonlinearity is of mild interest, although now we see that it is really rather an old friend amongst chemical systems. The real fun starts when the chemistry allows for *feedback* as well. Feedback arises when the products of later steps in the mechanism influence the rate of some of the earlier reactions (and, hence, the rate of their own production). This may take the form either of positive feedback (self-acceleration) or negative feedback (selfinhibition). Several examples of feedback and the consequences it may have on chemical systems are given in Table 1.3. We will consider some specific examples in the next section, but here indicate the general features.

All of the rate laws illustrated in Fig. 1.1, even the nonlinear ones, show rates decreasing monotonically as the extent of reaction increases. These systems have their maximum rate initially and as the reactants are consumed, so the rate falls. These are classed as *deceleratory reactions*. Self-accelerating or *acceleratory systems* are characterised by rate-extent curves that as well as being nonlinear, show an initial increase in rate with increasing extent of reaction. Some possible examples are indicated in Fig. 1.2. The maximum rate now occurs for some nonzero extent, once the reaction has really got itself going. Ultimately the rate begins to fall, tending to zero at complete conversion. Curves (a) and (b) show the two basic forms: an approximate parabola, with the rate increasing relatively quickly in the initial phase of the reaction and the maximum at roughly 50% conversion, and a 'cubic-type' curve that increases more slowly in the initial stages and attains its maximum nearer to the end of the reaction. The parabolic curve (a) can be usefully approximated by fitting a rate law of the form

$$d\xi/dt = k\xi(1 - \xi) \tag{1.10}$$

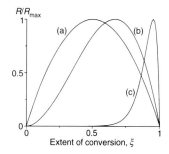

Fig. 1.2 The dependence of reaction rate $R = (-d\xi/dt)$ on extent of conversion ξ for example acceleratory systems with feedback: (a) quadratic autocatalysis; (b) cubic autocatalysis; and (c) an exothermic reaction. The rate is again normalized by the maximum rate R_{max} in each case, but this does not now occur at zero extent of reaction.

Table 1.3 Examples of feedback and its consequences in chemical systems

Gas phase (combustion processes, isomerizations)
chain branching	ignition limits
self-inhibition	oscillations and chaos
thermal feedback	thermal runaway
chain-thermal feedback	engine knock

Solution phase (BZ family, redox reactions)
autocatalysis	clock reactions
self-inhibition	travelling wavefronts and wave trains
	oscillations and chaos
	spatial patterns

Solid phase (pyrotechnics)
thermal feedback	nonsteady burning velocity
	thermal runaway (haystacks, etc.)

Gas–liquid (partial oxidation processes)
phase exchange	oscillations and chaos

Gas–solid (heterogeneous catalysis, catalytic convertors)
competitive adsorption	multistability
concentration dependent adsorption or phase changes	oscillations and chaos

Solid–liquid (electrodissolution, corrosion)
pH-dependent dissolution potentials	oscillations and chaos
film growth	

Biological systems (enzyme processes, glycolysis, nerve conduction, life)
allosteric effects	oscillations and chaos
chemotaxis	signal transmission
	spatial aggregation
	limb development
	evolution of form

and curve (b) by

$$d\xi/dt = k\xi^2(1-\xi) \tag{1.11}$$

These two rate laws have sometimes been translated into law of mass action forms such as

quadratic autocatalysis $\quad$ A + B $\to$ 2B $\qquad$ Rate = $k_q ab$ $\quad$ (1.12)

cubic autocatalysis $\quad$ A + 2B $\to$ 3B $\qquad$ Rate = $k_c ab^2$ $\quad$ (1.13)

where a and b represent the concentrations of A and B respectively. These are *model schemes*, in the sense described at the beginning of Section 1.1, for chemical autocatalysis. A more formal definition of autocatalysis involves the occurrence of a term in the rate equation for a given species in which that species appears with a positive stoichiometric coefficient. Thus, in the above models, $db/dt = +k_q ab$ or $+k_c ab^2$ with a stoichiometric coefficient of $+1$.

Many chemical systems can be modelled qualitatively by one or other of these forms, or by a weighted combination of the two. We will discuss the iodate–iodide reductant system which has such 'mixed' quadratic and cubic autocatalysis.

Also shown in Fig. 1.2 is the curve (c) for which $d\xi/dt = (1 - \xi)e^{B\xi}$. This approximates the thermal feedback for a first order, exothermic chemical reaction in an adiabatic reactor: the parameter B typically has a value of 20, making the curve peak at high extents of reaction.

1.3 Specific examples of chemical feedback

The examples of oscillations, travelling waves, etc., throughout this book will be based mainly on three specific chemical systems: the BZ reaction, the $H_2 + O_2$ reaction and the iodate–reductant reaction. Each of these has a mechanism for chemical feedback that can be exposed in relatively familiar kinetic arguments.

Belousov–Zhabotinsky reaction

The core of the feedback process in the BZ system is formed by the pair of overall steps

(8) $\qquad BrO_3^- + HBrO_2 + H^+ \overset{k_8}{\rightleftharpoons} 2BrO_2^\cdot + H_2O$

(9) $\qquad BrO_2^\cdot + M_{red} + H^+ \overset{k_9}{\rightarrow} HBrO_2 + M_{ox}$

Here M_{red} and M_{ox} are the reduced and oxidized forms, respectively, of a metal ion redox catalyst, such as the Ce(III)/Ce(IV) or ferroin/ferriin couples. So, one $HBrO_2$ forms two $BrO_2^\cdot$ radicals. If each of these go on to produce an $HBrO_2$ molecule in step (9) we have the overall stoichiometry

$$BrO_3^- + HBrO_2 + 3H^+ + 2M_{red} \rightarrow 2HBrO_2 + 2M_{ox} + H_2O \qquad (1.14)$$

which shows a strong similarity to the quadratic rate law (1.12), with $B = HBrO_2$.

The radical intermediate $BrO_2^\cdot$ is particularly reactive and tempts us to use one or other of the kineticists favourite tricks that reveal the actual mathematical form for the autocatalysis in these steps. First we write the net rate of production of $HBrO_2$ through this process:

$$d[HBrO_2]/dt = -k_8[BrO_3^-][HBrO_2][H^+] + $$
$$k_{-8}[BrO_2^\cdot]^2 + k_9[BrO_2^\cdot][M_{red}][H^+] \qquad (1.15)$$

If the Ce(III)/Ce(IV) couple is used, then reaction (9) is rather slow and reaction (8) rapidly attains a *quasi-equilibrium* balance. Thus

$$[BrO_2^\cdot]_{eq} = \{k_8[BrO_3^-][HBrO_2][H^+]/k_{-8}\}^{1/2} \qquad (1.16)$$

Substituting this into Eqn (1.15) then gives

$$d[HBrO_2]/dt = +k_9(k_8/k_{-8})^{1/2}[BrO_3^-]^{1/2}[M_{red}][H^+]^{1/2}[HBrO_2]^{1/2} \quad (1.17)$$

The positive sign indicates that this is indeed an autocatalytic form which is suitably represented by the overall process.

$$\tfrac{1}{2}HBrO_2 + \tfrac{1}{2}BrO_3^- + M_{red} + \tfrac{1}{2}H^+ \rightarrow HBrO_2 + M_{ox} + \tfrac{1}{2}H_2O \quad (1.18)$$

which is simply one half of Eqn (1.14) but here carries with it the implication of half-order kinetics with respect to the autocatalyst.

If the ferroin/ferriin couple is used, reaction (9) is relatively fast compared to the reverse step in reaction (8). In this case, a simple *steady-state* treatment for [BrO$_2^-$] neglecting the reverse step yields

$$[BrO_2^-]_{ss} = (2k_8/k_9)[BrO_3^-][HBrO_2]/[M_{red}] \quad (1.19)$$

Substituting this into Eqn (1.15) and, again, ignoring the reverse step (−8), we find

$$d[HBrO_2]/dt = +k_8[BrO_3^-][HBrO_2][H^+] \quad (1.20)$$

showing autocatalytic growth depending on the first power of the HBrO$_2$ concentration.

The H$_2$ + O$_2$ reaction

Feedback in gas-phase reactions has exactly the same form as autocatalysis, but is usually referred to as *chain branching*. The *branching cycle* that drives this reaction is formed from reactions (1)–(3) from the set given in Section 1.1. The easiest way to see this in operation is to imagine the fate of a single H atom entering this cycle. This reacts via step (2) to produce an OH radical and an O atom. The O atom reacts relatively quickly via step (3) to produce a second OH radical and to yield an H atom. The two OH radicals also both react quickly, via step (1), each yielding an H atom. Thus one H atom has given rise to three, a net gain of two radicals in the system. Taking 2(1) + (2) + (3) gives the overall stoichiometry

$$H + 3H_2 + O_2 \rightarrow 3H + 2H_2O \quad \text{Rate} = k_2[H][O_2] \quad (1.21)$$

The slowest, or *rate determining step* in this sequence is step (2), so the rate of increase of the H atom concentration grows with the increasing concentration of H atoms. We may also note, in passing, that the high free energy demand in effectively dissociating an H$_2$ molecule into two H atoms is met by the accompanying production of two molecules of water.

A simple steady-state analysis for the full scheme (0)–(4), setting d[O]/dt and d[OH]/dt = 0 in Table 1.2, yields the following equation for d[H]/dt:

$$d[H]/dt = 2k_0[H_2][O_2] + (2k_2[O_2] - k_4)[H] \quad (1.32)$$

The coefficient for [H] in the last term on the right-hand side measures the relative rate of the competing branching cycle and the termination step (4) that removes radicals from the system. This is known as the *net branching factor*, $\phi = 2k_2[O_2] - k_4$, and may be either positive or negative depending on the values of k_2, k_4, and [O$_2$]. The value of k_2 is particularly sensitive to the temperature as this step has a high activation energy. The values of k_4 and [O_2] are pressure-dependent: k_4 decreases as the pressure increases as radicals diffuse less easily to the walls, where termination occurs. Under conditions such that ϕ is positive, Eqn (1.32) indicates exponential growth in the radical concentration, leading to an ignition event. If termination wins, $\phi < 0$ and there is no net autocatalysis.

Iodate–iodide–reductant reaction

The two cases considered so far typically give rise to 'quadratic-type' autocatalysis. A higher-order dependence of the reaction rate on the autocatalytic species arises in the system of acidified iodate ions in the presence of a suitable reductant such as arsenite ion AsO_3^{3-}. The chemistry here is not so well established, but can be represented by two overall processes with empirical rate laws. First, there is the reduction of iodate by iodide to give iodine in the Dushman reaction:

(10) $\qquad IO_3^- + 5I^- + 6H^+ \rightarrow 3I_2 + 3H_2O$

The rate law for this process, $R_\alpha = -d[IO_3^-]/dt$, has been determined experimentally to involve the contribution of two channels that differ in their order with respect to iodide:

$$R_\alpha = (k_{\alpha 1} + k_{\alpha 2}[I^-])[I^-][IO_3^-][H^+]^2 \qquad (1.34)$$

with $k_{\alpha 1} = 4.5 \times 10^3$ M^{-3} s^{-1} and $k_{\alpha 2} = 1.0 \times 10^8$ M^{-4} s^{-1}.

The iodine thus formed is reduced back to iodide in a rapid process known as the Roebuck reaction

(11) $\qquad H_3AsO_3 + I_2 + H_2O \rightarrow H_3AsO_4 + 2I^- + 2H^+$

which has an empirical rate law,

$$R_\beta = -d[I_2]/dt = k_\beta [I_2][H_3AsO_3]/[I^-][H^+] \qquad (1.24)$$

with $k_\beta = 3.2 \times 10^{-2}$ M s^{-1}.

Combining (10) + 3(11), gives the overall stoichiometry

$$IO_3^- + 5I^- + 3H_3AsO_3 \rightarrow 6I^- + 3H_3AsO_4 \qquad (1.25)$$

so there is net production of I^- at a rate determined by the Dushman process:

$$d[I^-]/dt = + (k_{\alpha 1} + k_{\alpha 2}[I^-])[I^-][IO_3^-][H^+]^2 \qquad (1.26)$$

In buffered solution, [H^+] is constant and so this rate law compares exactly with the mixed quadratic and cubic form of Section 1.1, with A = IO_3^- and the autocatalyst B = I^-.

Another reductant, bisulphite ion HSO_3^-, can also be used in this reaction with similar effect. In this case, there is a slight modification as HSO_3^- reacts directly with IO_3^- to produce

(12) $\qquad IO_3^- + 3HSO_3^- \rightarrow I^- + 3HSO_4^-$

with

$$R_\gamma = -d[IO_3^-]/dt = k_\gamma[IO_3^-][HSO_3^-] \tag{1.27}$$

This system forms the classic Landolt clock reaction, widely used in lecture demonstrations, and which we will discuss in Chapter 2.

Table 1.3 lists some other forms of feedback arising in a range of physical situations and some of the resulting responses observed in consequence.

1.4 Closed and open systems

Closed systems

Many reactions are still carried out or occur naturally in thermodynamically closed systems. Some special considerations apply to such systems that limit the range of possible, long-time behaviour—although not as much as was once believed. Once a closed system has been assembled with specific initial concentrations, pressure and temperature, etc., there is a uniquely specified corresponding chemical equilibrium state. Multiple equilibrium solutions are not allowed. Not only this, but the final approach of the system to this equilibrium state cannot have any overshoot: reactions cannot oscillate about their chemical equilibrium state. This distinguishes chemical from mechanical systems such as a pendulum which can settle to its final state in a damped oscillatory manner.

This restriction is based on the principle of *detailed balance*. All chemical reactions are formally reversible to some greater or lesser extent. Thus we should write reaction (1.9) as

$$A + BC \rightleftharpoons AB + C \tag{1.28}$$

with the net rate of reaction then given by

$$-d[A]/dt = d[AB]/dt = k_+[A][BC] - k_-[AB][C] \tag{1.29}$$

where k_+ and k_- are the forward and reverse rate constants.

The stoichiometric equation for the $H_2 + O_2$ reaction can also be written as a reversible process

$$2H_2 + O_2 \rightleftharpoons 2H_2O \tag{1.30}$$

although this case is more complicated. Not only will the equilibrium composition have nonzero concentrations of the reactant and product species, there will also be nonzero concentrations of the various radical intermediates such as O, H, and OH. All of the elementary steps in the kinetic mechanism will be more or less reversible.

At the equilibrium state, the concentrations of the participating species adjust themselves so that the net rate of production of each species is zero simultaneously. The rate equation for each species will contain various production terms, corresponding to the reaction steps that produce the given species, and removal terms corresponding to steps that convert that species to another (these include the reverse of each production step). The equilibrium state is achieved not just by ensuring that the net rate of production exactly equals the net rate of removal for each species, but that for *every* reaction step in the mechanism *the forward rate becomes exactly equal to the reverse rate*: all the reaction steps are balanced. This is a very special requirement, appropriate only to chemical reactions in closed systems. (Perhaps it is not surprising that it needs infinite time for the equilibrium state to be attained.)

The requirement of a monotonic approach to the chemical equilibrium state does not, however, rule out 'interesting behaviour' such as oscillations in closed systems. Reactions are usually set up with initial conditions that are far from the corresponding equilibrium state, and chemical systems 'far from equilibrium' are not constrained so tightly. It does, however, mean that such oscillations cannot last forever: they must cease as the system gets close to its equilibrium state, i.e., after a sufficiently long time. Oscillations are necessarily transient, but possibly very long-lived, in closed systems.

Open systems

When systems exchange mass with their surroundings, and hence are thermodynamically *open*, the inflow and outflow means that individual molecules spend only a finite time in the reactor. The simplest example of an open system is the continuous-flow reactor, particularly if this is well stirred. More complex examples include living things and atmospheres. The finite *residence time* in such systems means that the reaction does not approach the chemical equilibrium state it would find if it were closed. The restriction mentioned above do not, therefore, apply.

In flow systems, reactions may evolve to *steady states* for which the rates of chemical production and removal of each species just balance the rates of inflow and outflow of those species. In some cases, more than one steady state may coexist. Oscillations may occur and last forever, and have the time to develop into more complex responses, as we shall see.

2 Clocks and fronts

The first two types of 'exotic behaviour' arising from chemical feedback that we will consider are the related phenomena of *clock reactions* and *chemical waves*. The iodate–arsenite and iodate–bisulphate mentioned in Section 1.3 provide a good introduction, but the phenomena are observed in a much wider range of systems from gas to solid phases and into biology.

2.1 Clock reactions

The classic clock reaction, discovered by Landolt in 1886, involves the autocatalytic iodate–bisulphite system. The reactants, in aqueous solution and with added starch indicator, are colourless. There is a long induction period, during which the iodide ion concentration increases slowly, followed by a rapid acceleration in the rate which leads to a sharp colour change. The effect is most spectacular and draws a gasp from even the most sophisticated audience. If the reductant is in stoichiometric excess, the colour fades again.

The induction period t_{ind} depends on the initial concentrations of the reactants iodate and reductant:

$$t_{ind} = k/[IO_3^-]_o[HSO_3^-]_o \qquad (2.1)$$

with $k = 4 \times 10^{-3}$ M²s. A solution that is initially 0.01M in each of the reactants has a 40s induction time. By varying the initial concentrations, the clock time can be adjusted over a wide range. The clock works equally well with arsenite as the reductant, although iodide ion normally then has to be added initially. The induction period is extremely sensitive to the initial concentrations of the iodide autocatalyst.

The basic origin of this clock behaviour can be discerned with reference back to Fig. 1.2. The Landolt system is qualitatively well represented by the cubic curve (b). At the beginning of the reaction, at $\xi = 0$, the rate is virtually zero because of the low iodide concentration. The rate at which the extent of conversion increases, and hence at which the system moves from left to right along the ξ-axis, is determined by the height of the curve. Thus the system crawls only very slowly away from the origin, with the rate only increasing slowly at first. Once the reaction has had sufficient time to develop, however, the rate begins to increase dramatically and so the progress along the ξ-axis accelerates.

The iodate–reductant system can be simulated successfully using the empirical rate-law forms presented in Section 1.3. Figure 2.1 shows the evolution of the reactant IO_3^-, autocatalyst I^- and intermediate I_2 for an iodate–arsenite system with excess reductant. The clock has an induction

---

The Landolt clock can be demonstrated by mixing 0.02M solutions of KIO_3 (2.14 g in 0.5 dm³) and $NaHSO_3$ (0.95 g of $Na_2S_2O_5$ in 0.5 dm³); 50 cm³ of each mixed with a few drops of 1% starch solution gives a clock time of approximately 1 minute. This can be lengthened by dilution. The bisulphite solution oxidizes in air over the course of several hours, so should not be prepared too far in advance of use.

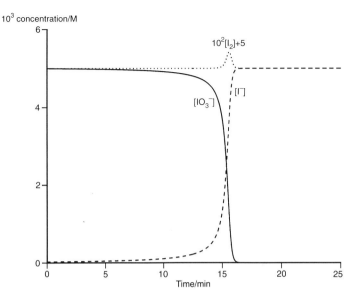

Fig. 2.1 The evolution of the concentrations of the reactant $[IO_3^-]$, product $[I^-]$ and intermediate $[I_2]$ for the Landolt clock reaction with initial concentrations $[IO_3^-]_o = 5 \times 10^{-3}$ M and $[I^-]_o = 2.5 \times 10^{-5}$ M and with $[H^+] = 7.1 \times 10^{-3}$ M.

period of c. 15 minutes. With excess iodate, I_2 remains after the reaction event and the iodide concentration shows a peak before falling back to zero.

This computation can readily be repeated for different initial concentrations or pH, as indicated in Fig. 2.2. At the expense of some approximation, however, the functional dependence of t_{ind} on $[IO_3^-]_o$, etc., can be obtained analytically. The cubic autocatalysis model, Eqn (1.13) was introduced in Section 1.2 and its similarity to the empirical rate law above, with $A = IO_3^-$ and $B = I^-$, mentioned in Section 1.4. If we assume that the Dushman process is rate-determining, that $[H^+]$ = constant and that the quadratic channel is not significant, then the iodate–reductant system can be approximated by the cubic form:

$$da/dt = -k_c ab^2 = -k_c a(a_o + b_o - a)^2 \qquad (2.2)$$

The second form here uses the condition $a_o + b_o = a + b$, or $[IO_3^-]_o + [I^-]_o = [IO_3^-] + [I^-]$, which is simply the conservation of iodine atoms, assuming that the concentration of I_2 remains very low. This conservation condition is then used to eliminate $[I^-]$, so the right-hand side of the rate law can be expressed solely in terms of the iodate ion concentration. The rate constant k_c is given by $k_c = k_{a2}[H^+]^2$.

Equation (2.2) can be integrated explicitly to yield

$$t = \frac{1}{k_c(a_o + b_o)^2} \left[\ln\left(\frac{a_o + b_o - a}{b_o}\right) + \frac{(a_o + b_o)(a_o - a)}{b_o(a_o + b_o - a)} \right] \qquad (2.3)$$

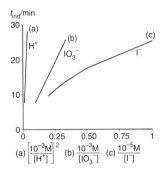

Fig. 2.2 The variation of the clock induction time with initial concentration of the reactants: (a) variation with $[H^+]$, with $[IO_3^-]_o = 5 \times 10^{-3}$ M and $[I^-]_o = 2.5 \times 10^{-5}$ M; (b) variation with $[IO_3^-]_o$, with $[H^+] = 7.1 \times 10^{-3}$ M and $[I^-]_o = 2.5 \times 10^{-5}$ M; (c) variation with $[I^-]_o$, with $[H^+] = 7.1 \times 10^{-3}$ M and $[IO_3^-]_o = 5 \times 10^{-3}$ M.

This allows the time for the iodate concentration to fall from its initial value a_o to any other value a to be calculated. The computed results in

Fig. 2.1 suggest that the colour change occurs when $a = \frac{1}{2} a_o$, i.e., when half the original iodate has reacted. If we also neglect the term b_o as small compared with a_o or $\frac{1}{2} a_o$ where possible, and note that the logarithmic term will generally provide only a small correction to the second term in the square brackets, we obtain the much simpler expression

$$t_{\text{ind}} = \frac{1}{k_c a_o b_o} = \frac{1}{k_{\alpha 2} [\text{H}^+]^2 [\text{IO}_3^-]_o [\text{I}^-]_o} \tag{2.4}$$

Thus the induction period should lengthen inversely with the initial iodate and iodide concentrations and as the inverse square of the H^+ concentration. These predictions are borne out in Fig. 2.2, although the dependence on $[\text{I}^-]_o^{-1}$ deviates from linearity at high initial iodide concentrations. Substituting in the data from Fig. 2.1, Eqn (2.4) predicts $t_{\text{ind}} = 27$ minutes. In fact the quadratic channel is not negligible under these conditions, at least in the initial stages, so agreement to within better than a factor of 2 cannot be expected. Nevertheless, the iodate–arsenite system does indeed show the above functional dependence on $[\text{H}^+]$, $[\text{IO}_3^-]_o$, and $[\text{I}^-]_o$ experimentally. We may also compare Eqn (2.4) with the empirical form (2.1). The latter applies to systems in which no H^+ is added other than as HSO_3^-. If the dissociation constant K_a for the process

$$\text{HSO}_3^- \rightleftharpoons \text{H}^+ + \text{SO}_3^{2-}, \quad K_a = [\text{H}^+][\text{SO}_3^{2-}]/[\text{HSO}_3^-] \tag{2.5}$$

is small compared with $[\text{HSO}_3^-]_o$, then a simple calculation shows $[\text{H}^+]^2 \approx K_a[\text{HSO}_3^-]_o$. Provided $[\text{I}^-]_o$ remains constant in a sequence of experiments, then the two expressions are consistent.

Other examples of clock reactions

Clock-type behaviour is observed not only in solution phase reactions, but also in a wider context. In gas phase reactions involving the oxidation of simple hydrocarbon such as n-butane, long induction periods, perhaps of the order of days, are also frequently observed. Chemical autocatalysis, arising from chain branching, is again involved but thermal feedback can also play a part. When heat transfer from the reactor is sufficiently slow that the heat released by the reaction causes significant self-heating of the gaseous reactants, the sharpness of the final acceleration increases dramatically. As the reaction typically is chemiluminescent (via emission from electronically excited HCHO), a brief flash of light can be seen just as the reaction goes rapidly from low extents of conversion to completion at the end of the induction period. This behaviour is known as the *pic d'arrêt*.

Even longer induction periods arise in the clock reactions associated with thermal runaway. The most well-known example of such *thermal explosions* is the apparently sudden ignition of haystacks, possibly many months after assembly. In fact, this is a relatively common phenomenon, arising wherever cellulosic material, coal or many other oxidizable 'fuels' are stored for extended periods in air.

The underlying cause of thermal runaway can be illustrated with a simpler model. We can imagine a pile of some reactant surrounded by a reservoir at some fixed ambient temperature T_a, as sketched in Fig. 2.3. An exothermic reaction proceeds at some low, but nonzero rate. The resulting heat released causes heating of the reactant to some higher temperature T. For simplicity, the temperature within the reactant will be taken to be uniform, as though the system were well stirred. Now, two effects must be considered. First, the reaction rate, and hence the rate of heat evolution, will increase because of the raised local temperature. Secondly, as the material is hotter than its surroundings, there will be heat transfer to the reservoir, perhaps by Newtonian cooling so the rate of heat loss depends on the temperature difference $\Delta T = T - T_a$. Depending on the size and shape of the pile, the ambient temperature and the exothermicity of the reaction, either the rate of heat loss will eventually grow to balance the rate of chemical heat evolution, giving rise to only small temperature differences, or the rate of heat evolution will always remain greater than that for heat transfer, in which case the temperature will continue to rise until and ignition occurs.

These two situations can be illustrated graphically by means of a *thermal diagram*. The rate at which the temperature within the pile changes with time is governed by the *heat balance equation*, which can be written in the form

$$c_p \sigma \frac{dT}{dt} = \underbrace{QcAe^{-E/RT}}_{\mathscr{R}(T)} - \underbrace{(\chi S/V)(T - T_a)}_{\mathscr{L}(T)} \qquad (2.6)$$

where c_p is the specific heat capacity and σ the density. The two terms on the right-hand side, $\mathscr{R}(T)$ and $\mathscr{L}(T)$ are the rates of chemical heat release and heat transfer per unit volume. In the heat release term, Q is the reaction exothermicity ($= -\Delta H$), c represents the 'concentration' of the reactant, whilst the group $Ae^{-E/RT}$ is the reaction rate constant written to show explicitly its temperature dependence. If the concentration term is treated as a constant, $\mathscr{R}$ varies with the temperature inside the pile as shown in Fig. 2.4, following the Arrhenius form. The heat transfer term $\mathscr{L}$ shows a linear dependence on the temperature rise, as indicated in Fig. 2.4. The slope is governed by $\chi S/V$, where χ is the heat transfer coefficient and S/V is the surface to volume ratio. Thus, the slope of the heat-loss term is determined by factors such as the size and shape of the pile. The intercept of $\mathscr{L}$ on the temperature axis is given by the ambient temperature, T_a.

Two heat-loss lines are shown in Fig. 2.4, corresponding to slightly different ambient temperatures. For the lower ambient temperature, $T_{a,1}$, the heat generation and heat-loss lines intersect at some temperature T_{ss} relatively close to $T_{a,1}$. This intersection represents a point of thermal balance, so $\mathscr{R}(T_{ss}) = \mathscr{L}(T_{ss})$ and $dT/dt = 0$. For this case, T will increase from T_a and approach this intersection temperature exponentially. A small steady-state temperature excess will be established. This represents *subcritical* behaviour. The case with ambient temperature $T_{a,2}$ is *supercritical*. $\mathscr{R}$ and $\mathscr{L}$ never intersect: the rate of heat evolution always exceeds the rate of heat loss, so dT/dt is always positive and the temperature within the pile increases steadily with time. At any given time, the rate at which the temperature is

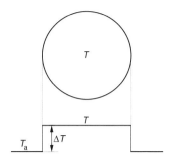

Fig. 2.3 Schematic representation of the Semenov model for a self-heating chemical system. The surroundings are held at a fixed temperature, but exothermic reaction increases the temperature within the reaction volume V to T. Heat transfer occurs by Newtonian cooling across the surface area A at a rate determined by the temperature excess $\Delta T = T - T_a$ and the heat transfer coefficient χ.

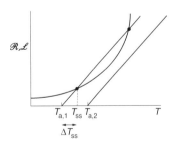

Fig. 2.4 Thermal diagram showing dependence of the rate of heat release $\mathscr{R}$ on temperature for the Semenov model. Also shown are two heat loss lines corresponding to two different ambient temperatures, one subcritical and one supercritical.

There is a difference between simple Landolt clock behaviour and thermal runaway: in the latter there is also a critical condition, with clock behaviour only if $T_a > T_{a,cr}$.

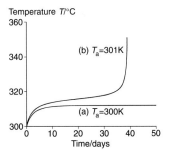

Fig. 2.5 Examples of temperature evolution within self-heating systems: (a) $T_a = 300$ K, subcritical evolution to a small steady-state temperature rise, $\Delta T_{ss} \approx 10$ K; (b) $T_a = 301$ K, supercritical thermal runaway after an induction period of approximately 38 days.

I. Nagypal and I. R. Epstein (1988). *J. Chem. Phys.*, **89**, 6925.

Recipes for travelling waves: T. A. Gribschaw, K. Showalter, D. L. Banville, and I. R. Epstein (1981). *J. Phys. Chem.*, **85**, 2152. (a) Excess iodate: $[IO_3^-]_0 = 0.03$M, $[H_3AsO_3]_0 = 0.05$M, pH = 1.80; (b) excess reductant: $[IO_3^-]_0 = 0.03$M, $[H_3AsO_3]_0 = 0.094$M, pH = 1.50, each with starch indicator.

increasing is proportional to the difference between $\mathfrak{R}$ and $\mathfrak{L}$. Because, the two curves almost touch at one point, this difference becomes very small during part of this evolution. The system has to crawl through the narrow channel between $\mathfrak{R}$ and $\mathfrak{L}$, during which time dT/dt is almost zero. The resulting temperature–time history is sketched in Fig. 2.5, indicating an initial rise in the temperature difference ΔT to a value very similar to the steady-state rise appropriate to the subcritical case, followed by a long induction period before a final acceleration into the ignition event.

Clock reaction behaviour has also been observed in the polymerization of a mutant form of haemoglobin, associated with the disease sickle cell anaemia, to form a highly viscous gel. The apparent rate law here indicates a dependence of rate on somewhere between the 20th and 30th power of the initial haemoglobin concentration.

Supercatalysis

Clock reaction behaviour is also observed in the reaction between chlorite and thiosulfate ions. In this case, however, the clock induction time appears not to be repeatable from experiment to experiment. Instead, the observed clock times are distributed statistically about a mean value, with the distribution and mean being reproducible from one series of experiments to another. The origin of this phenomenon is not completely clear, but has been termed *supercatalysis* to indicate an especially high degree of chemical feedback through the autocatalytic species H^+. In fact, the formal kinetic order of this reaction is approximately 10, compared with the order of 5 for the Landolt system. A 1% error in dilution between successive experiments would thus give rise to a factor of $(1.01)^{10} = 1.1$, i.e., of approximately 10% in the initial rate and may be expected to have similar implications on the observed clock time. The sickle cell haemoglobin clock also shows such a distribution of clock times.

2.2 Fronts

In the Landolt clock reaction, the reactants are usually well-mixed initially. Any addition of the autocatalyst I^- is usually done with a swirl of the mixture to ensure an even distribution, so all parts develop at the same rate towards the final, equilibrium state. As a variation on this theme, the initiation may be performed in a different way, confined to a small localized region. For instance, reaction may be initiated at a point in a thin layer in a petri dish using a thin Pt electrode negatively biased with respect to the counter-electrode. This causes a local increase in $[I^-]$ via electrochemical reduction of IO_3^-. More simply, a small drop of relatively concentrated iodide solution can be added to the top of a test-tube containing the iodate–reductant mixture. In these cases, the clock reaction spreads away from the initiation site, giving rise to a spatially resolved clock or *chemical wavefront*. Some experimental examples are shown in Fig. 2.6.

The iodate–arsenite system is particularly suited to this phenomenon: there is virtually no spontaneous production of I^- in the absence of the

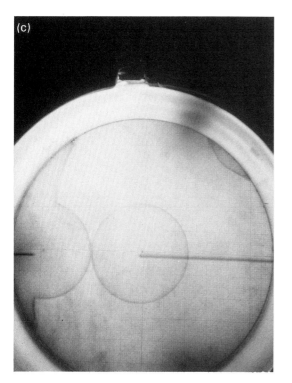

Fig. 2.6 Experimental examples of travelling waves in the iodate–arsenite reaction: (a) iodate is in excess so the region after the wave is blue; (b) excess arsenite so region after wavefront is colourless (wave propagating downwards); (c) circular wave propagating outwards from central initiation by wire electrode, excess arsenite (K. Showalter, with permission. See also A. Hanna, A. Saul and K. Showalter (1982). *J. Am. Chem. Soc.* **104**, 3838–44).

autocatalyst, especially if the impurity traces of iodide are removed from the iodate stock solution by careful recrystallization or by addition of a precipitating agent such as Cd^{2+}. The reactants ahead of any wavefront are thus effectively 'kinetically frozen' in their initial state and rely on the diffusion of iodide from the wave to initiate the conversion to products locally. Once the reaction begins, locally, much more iodide is produced and this can then diffuse into neighbouring regions, initiating the clock reaction there. Thus, the wave is a combined reaction-diffusion process that converts the system from its initial state to its final equilibrium composition. This is the characteristic feature of a *front*. Other types of wave (*pulses* and *wavetrains*) will be introduced in Chapter 4.

The evolution of the iodide concentration and the development of a pair of reaction-diffusion fronts is illustrated in Fig. 2.7. A small input of the autocatalyst I^- is introduced close to the origin at $t = 0$. Reaction is initiated and the autocatalyst begins to spread: the initial 'top hat' grows and widens. Eventually, all the iodate is consumed in the vicinity of the initiation site. The local concentration of iodide thus established slightly exceeds the initial iodate concentration, by an amount that depends on the initial input of iodide: this excess slowly decays by purely diffusional spreading over a longer timescale. The final three concentration profiles in Fig. 2.7 reveal the outward propagating fronts. Ahead of each front is a region of unreacted iodate and no iodide: behind each front the composition corresponds to complete reaction. These profiles have been computed at equal time intervals. This reveals another characteristic feature of such reaction-diffusion fronts: *they travel at a constant velocity through the reaction mixture*. Typical wave speeds for the iodate–arsenite reaction are of the order of 1 mm min^{-1}. We may compare this with the timescale for simple diffusional spreading. For a simple one-dimensional configuration, the mean displacement due to diffusion after time t is given by $\bar{x} = 2(Dt)^{1/2}$, where D is the diffusion coefficient. Typical values for D for small ions in aqueous solutions are of the order of 10^{-5} cm^2 s^{-1}, thus the time taken for diffusion to cause spreading over 1 cm would be approximately 400 min. This is more than an order of magnitude slower than the progress of the reaction-diffusion front, which would cover 1 cm in approximately 10 min. The enhancement in velocity arises because only small amounts of the autocatalyst have to diffuse before the chemistry is stimulated into producing more. Reaction-diffusion processes are believed to underlie the signalling mechanisms in many biological systems which exploit this potential for enhanced propagation velocity.

To examine the propagation of these fronts, and to derive the relationship between the wave-speed and the kinetics, we can again use the simple cubic autocatalysis model Eqn (1.13). Somewhat fortunately, as well as being a reasonable approximation to the iodate–arsenite system, it also provides analytical results rather than requiring numerical solution.

If we imagine a long thin tube, so that we need only consider diffusion in the x-coordinate, the reaction-diffusion equations governing the rates of change of the iodate and iodide concentrations, a and b, are

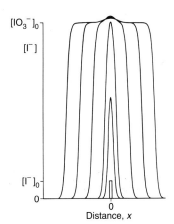

Fig. 2.7 Typical evolution of a travelling wave: iodide ion concentration developing with time for the Landolt system from an initial localized input $[I^-]_0$ at $t = 0$. The profiles are presented at equal time intervals, indicating the evolution of a constant wave-speed.

$$\partial a/\partial t = D_A \partial^2 a/\partial x^2 - k_c ab^2 \quad (2.7\text{a})$$

$$\partial b/\partial t = D_B \partial^2 b/\partial x^2 + k_c ab^2 \quad (2.7\text{b})$$

Here the terms $D_A \partial^2 a/\partial x^2$ and $D_B \partial^2 b/\partial x^2$ are Fick's Law for the diffusion process, with D_A and D_B being the diffusion coefficients for the reactant and autocatalyst, and $k_c = k_{a2}[H^+]^2$ as before. If the wavefront is travelling from left to right, then behind the wave $(x \to -\infty)$ $a = 0$ and $b = a_o$, where a_o is the initial concentration of iodate, corresponding to the fully reacted state. Ahead of the wave $(x \to +\infty)$, $a = a_o$ and $b = 0$. We can use the conservation equation for iodine atoms, which here is $a_o = a + b$ as $b_o = 0$ in the bulk of the solution, to eliminate one of the concentrations. Thus, working with Eqn (2.7a):

$$\partial a/\partial t = D\partial^2 a/\partial x^2 - k_c a(a_o - a)^2 \quad (2.8)$$

This conservation condition assumes that A and B have equal diffusion coefficients: $D_A = D_B = D$. This is typically the case for solution phase reactions. If D_A and D_B are not equal, a greater variety of chemical wave behaviour is possible: see Section 2.3.

This equation has an analytical solution, giving the concentration of iodate or iodide as a function of position and time, that can be written as

$$a(x,t) = a_o - b(x,t) = a_o/\{1 + \exp[-(x - ct)/l_d]\} \quad (2.9)$$

The reaction front propagates at a constant velocity c, given by:

$$c = \{\tfrac{1}{2} D k_c a_o^2\}^{1/2} \quad (2.10)$$

Reaction is virtually confined to a narrow region in which both A and B have significant concentrations. The thickness of this zone is related to the length scale l_d:

$$l_d = (2D/k_c a_o^2)^{1/2} \quad (2.11)$$

Equation (2.10) is of considerable significance. In terms of the iodate–arsenite system, the wave speed is given by

$$c = \{\tfrac{1}{2} D k_{\alpha 2}[IO_3^-]_o^2 [H^+]^2\}^{1/2} \quad (2.12)$$

There are three factors involved under the square root sign: a simple numerical factor of $\tfrac{1}{2}$, the diffusion coefficient D and a pseudo-first-order rate constant $k_{\alpha 2}[IO_3^-]_o^2[H^+]^2$ for the autocatalytic step. This is, in fact, a general functional form. For the quadratic autocatalysis model Eqn (1.20), the corresponding result is

$$c = 2\{Dk_q a_o\}^{1/2} \quad (2.13)$$

so only the numerical factor has change from $\sqrt{\tfrac{1}{2}}$ to 2: the dependence on the square root of the diffusion coefficient and the rate constant for the autocatalytic process remains. Systems which exhibit waves following quadratic-type autocatalysis include the bromate– and nitrate–ferroin reactions. Quadratic feedback also occurs in population models, such as the Fisher–Kolmogorov equation, and in SIR (susceptible-infected-removed) models for the spread of infectious diseases such as AIDS or rabies.

Pushing this comparison further, flames in exothermic combustion systems are essentially reaction-diffusion fronts, travelling at constant speeds. The role of the autocatalyst is played by temperature rise resulting from the chemical heat release with the feedback acting through the Arrhenius

The equations for a 'one-dimensional, premixed, laminar flame' can be written as

$$\partial a/\partial t = D\partial^2 a/\partial x^2 - k(T)a$$
$$\partial T/\partial t = (\kappa/c_p\sigma)\partial^2 T/\partial x^2 + (q/c_p\sigma)k(T)a$$

where q is the reaction exothermicity ($-\Delta H$). The group ($q/c_p\sigma$) gives the temperature rise ΔT_{ad} in the system due to complete reaction under adiabatic conditions. These have a similar form to eqns (2.6a and b), with T replacing b. The dependence of the reaction rate term on T is given by the Arrhenius form $k(T) = A\exp\{-E/RT\}$, replacing the b^2 term in autocatalytic equations.

temperature dependence of the rate constant. The flame speed can also be expressed in the above form:

$$c = \{2(\kappa/c_p\sigma)k(T_b)/B^2\}^{1/2} \tag{2.14}$$

Here the group $\kappa/c_p\sigma$, where κ is the thermal conductivity, is known as the thermal diffusivity; $k(T_b)$ is the reaction rate constant evaluated at the temperature of the burnt gas and B is a numerical factor related to the temperature rise of the burnt gas over the cold, unheated reactants far ahead of the flame. Equation (2.14) again has the propagation speed proportional to the square root of a diffusivity and a characteristic rate constant. Typical flame speeds are of the order of 0.1 to 1 ms^{-1}.

Returning to the iodate–arsenite system, Eqn (2.12) predicts that the wave-speed will increase linearly with both the initial iodate concentration and with [H$^+$]. The front thickness, however, will vary inversely with these concentrations, from Eqn (2.11). For a system with [IO$_3^-$]$_o$ = 5 × 10^{-3} M, [H$^+$] = 7.1 × 10^{-3} M and D = 2 × 10^{-5} cm^{2}s^{-1}, we have c = 0.66 mm min^{-1} and l_d = 0.18 mm. If the wave-speed and thickness can be measured in the same system, they provide a route to the experimental determination of the reaction rate constant and the diffusion coefficient of the autocatalytic species.

For the circular wave initiated from a thin wire electrode and propagating outward in a petri dish, the initial wave-speed is usually slower than that predicted above (which effectively applies to a planar front). The influence of *curvature* on wave-speeds will be particularly important when we consider different types of chemical waves in Chapter 4. A curved wavefront allows the autocatalyst diffusing ahead to become diluted by simple geometric spreading, thus reducing the chemical reaction rate, leading to a lower propagation speed. As the radius of the circular wave increases, however, the curvature decreases and eventually the wave front appears locally to be effectively planar, and the above wave-speed is achieved.

2.3 Oscillatory and nonplanar fronts

The simple one-dimensional wavefronts described above, propagating with constant speed and shape, arise for systems in which the reactant and autocatalytic species diffuse with roughly the same diffusion coefficient. In the case of nonisothermal flames, the equivalent case is that for which the molecular and thermal diffusivities are approximately equal. These are not unusual situations. However, there are also cases of interest in which the reactant and feedback (autocatalyst or heat) species diffuse at significantly different rates. This allows some additional phenomena.

For isothermal systems, the reactant and autocatalyst may have different mobilities if they have very different molecular sizes. The same effect arises if one of the species is immobilized by adsorption onto a solid phase (as occurs in chromatography or with ion-exchange resins, for example) or forms a complex with a larger ligand, especially if that ligand can be immobilized in a suitable gel. In the Landolt system, iodide complexes with cyclodextrin are distributed through a polyacrylamide gel to effectively reduce its diffusivity: in the BZ system, the positively-charged catalyst can be adsorbed into a cation-exchange resin whilst the anionic reactant species BrO_3^- remains free to diffuse. For flame systems with light fuels such as H_2, the molecular diffusivity is significantly higher than the thermal diffusivity: for heavier hydrocarbon fuels and for solid phase pyrotechnic systems, in which the fuel is effectively immobile, the opposite situation arises.

Feedback species with higher diffusivity: oscillatory wave speed

If the thermal diffusivity is sufficiently higher than the molecular diffusivity and the reaction is sufficiently exothermic, the premixed laminar flame described above can cease to support a combustion front propagating at a steady speed and constant shape. Instead, the speed of the flame oscillates in time, with the flame constantly accelerating and decelerating. The flame never actually goes backwards (into the burnt products) but its forward speed varies continuously as it propagates into the unburnt reactants. The shape of the flame front also varies in time, as indicated in Fig. 2.8. For some of the oscillatory period, the local temperature in the flame front can exceed the equilibrium adiabatic temperature rise. Whilst this local 'superheating' is occurring, the flame advances more rapidly, driven by the enhanced diffusion of heat into the region ahead. However, this extra heat transfer enhances the reaction rate which leads to a depletion of the reactant concentration ahead of the flame. The development of a maximum in the temperature profile also leads to an additional heat loss process, cooling by conduction to the less hot products behind as well as to the cold reactant ahead. The superadiabatic temperatures cannot be sustained. The front slows as it passes through the depleted region before entering a region in which the reactant concentration remains high and the acceleration process and superadiabatic temperatures can develop again. Oscillating fronts of this nature have been observed experimentally in solid ('gasless') pyrotechnic mixtures. Oscillatory fronts are also supported by higher orders of autocatalysis if the diffusion coefficient of the reactant is sufficiently small.

Feedback species with lower diffusivity: patterned fronts

For the opposite case in which the reactant diffuses more rapidly than the autocatalyst or heat, the significant changes arise when waves propagating in two or three dimensions are considered. The important qualitative change in this case can be illustrated by considering propagation of a chemical front along a long, thin two-dimensional strip. The analogue of the one-dimensional fronts discussed above would here be a planar wave propagating in the x-direction so that there are no concentration gradients in the y-direction. However, if the diffusivities are sufficiently different and the feedback is sufficiently strong (cubic rather than quadratic), such planar fronts may become spontaneously unstable and a 'wiggly' front emerges. Figure 2.9 indicates the development of a nonplanar front for a case of cubic autocatalysis in which $D_A/D_B = 5$ (the Landolt system with $D_{IO_3^-}/D_{I^-} = 5$).

D. Horvath, V. Petrov, S. K. Scott, and K. Showalter (1993). *J. Chem. Phys.*, **98**, 6332–43; K. Showalter and S. K. Scott (1992). *J. Phys. Chem.*, **96**, 8702–11

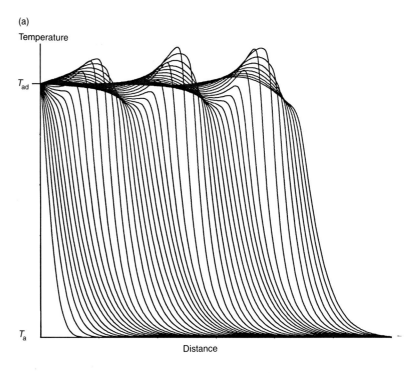

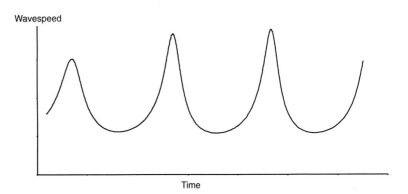

Fig. 2.8 Nonsteady combustion: the propagation of a flame with an oscillating velocity: (a) variation of shape; (b) variation in instantaneous velocity.

The mechanism for this instability can be described in qualitative terms. Relative to a planar front, a portion of a nonplanar front in the autocatalyst concentration that is advanced and hence convex allows an enhanced geometrical spreading of the autocatalyst into the reactant ahead of the front zone, as indicated in Fig. 2.10. This dilutes the autocatalyst concentration and so tends to reduce the local speed. However, the same convexity allows an enhanced diffusion of reactant into the reaction zone, tending to increase the local front speed. In the situation where $D_A > D_B$, the second

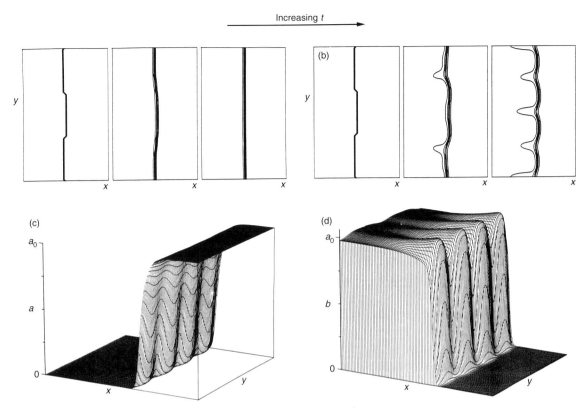

Fig. 2.9 The response of a planar two-dimensional wave to spatial perturbation: (a) $D_A = D_B$, the planar front is stable and the perturbation decays; (b) $D_A = 5D_B$, the planar front is unstable and a nonplanar form emerges. The contours on these diagrams represent lines of equal concentration with $b = 0.95, 0.75, 0.55, 0.35$ and $0.15 a_0$. The full a and b concentration profiles, (c) $[IO_3^-]$ and (d) $[I^-]$ for the Landolt reaction, corresponding to the longest time are also shown.

effect wins out and so the local speed is increased relative to that of a planar wave. This will tend to increase the degree of advancement of the convex segment of the front. Conversely, at a retarded segment, there is a geometrical concentrating effect on the diffusion of the autocatalyst, but a geometrical dilution of the reactant. If the latter is more significant, the local wave speed is reduced relative to that of a planar wave, leading to further retardation of that segment of the front. These discussions show that any pertubation of the planar front can be reinforced by the system so that nonplanar fronts are established. This mechanism also lies behind the development of the cellular structure of many hydrocarbon flames. The same qualitative arguments show that if $D_B > D_A$, then the planar wave is stabilized to such spatial pertubations. Other types of chemical pattern formation will be discussed in Section 4.9.

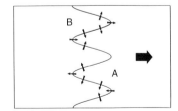

Fig. 2.10 Geometrical effects for a nonplanar front.

3 Resetting the clock: oscillations

Clock reactions and single travelling wavefronts typically arise as one-off reaction events, taking the initial reactants through to the final products. For repetitive events, such as an oscillatory reaction in a well-stirred system, there must be some way of resetting the clock.

3.1 The BZ reaction

Recipe for BZ reaction: 500 ml 1M H_2SO_4 (if this is made up by diluting 27 ml of concentrated acid just beforehand, the warm mixture oscillates at a more convenient rate), 14.3 g malonic acid, 5.22 g $KBrO_3$ and 0.548 g $Ce(NH_4)_2(NO_3)_6$ with 1–2 ml of 0.025M ferroin indicator (1.485 g 1:10 phenanthroline and 0.695 g $FeSO_4.7H_2O$ in 100 ml H_2O). Pour some of this mixture as a thin layer in a petri dish to see chemical waves develop (Chapter 4).

The most familiar example of an oscillatory chemical reaction is the BZ system. This reaction involves the oxidation of an organic species such as malonic acid by an acidified bromate solution in the presence of a metal ion catalyst. Various metal ions can be employed, with the Ce(III)/Ce(IV) and $[Fe(II)(phen)]^{2+}/[Fe(III)(phen)]^{3+}$ (ferroin/ferriin) couples most widely used. In a closed (batch) system, the reaction typically exhibits a short induction period, followed by an oscillatory phase. The colour alternates between red and blue (for the ferroin/ferriin couple) with a period of approximately one minute. The oscillations may last for over two hours during which perhaps a hundred oscillations are observed. Ultimately, the oscillations die out and the system then drifts slowly and monotonically towards it chemical equilibrium state. Typical experimental records, as measured by Pt and Br^--sensitive electrodes, each referenced to a calomel electrode, are shown in Fig. 3.1. The Pt electrode responds primarily to the metal ion redox couple: this shows a sharp change from the reduced to the oxidized state followed by a more gradual return. The sharp switch is also associated with the abrupt colour change. The bromide electrode responds primarily to $[Br^-]$. A slow decrease AB in $[Br^-]$ can be observed, before the sharp drop BC that accompanies the oxidation of the metal catalyst and the colour change. This gives rise to a *relaxation waveform*, with a second segment of relatively slow evolution CD before the bromide ion concentration increases rapidly again DA.

Although there are oscillatory variations in the concentrations of some 'intermediate' species, it is also important to note that the concentrations of the major reactants, bromate and malonic acid, decrease slowly but continuously during the reaction process (stepwise, during the oscillatory phase). In consequence, the reaction continuously flows in the direction of decreasing free energy: there is no oscillation in the direction of the

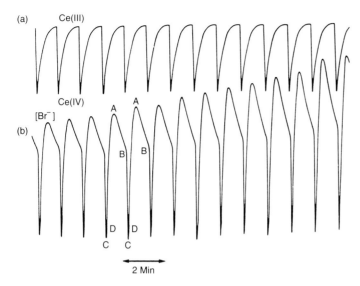

Fig. 3.1 Typical experimental records for the BZ reaction in a closed vessel showing oscillations in the potentials of (a) Pt electrode and (b) Br⁻ electrode, both referenced to calomel.

overall reaction which is always moving inexorably towards the chemical equilibrium state.

3.2 Mechanism for BZ reaction

Inorganic clock reaction

The understanding of the BZ system has been developed primarily in terms of the Field–Körös–Noyes (FKN) mechanism. The major feedback routes arise through Process A, which removes bromide ion, and Process B, which provides the autocatalysis.

R.J. Field, E. Körös, and R.M. Noyes (1972). *J. Am. Chem. Soc.*, **94**, 8649–64.

The important steps in Process A are:

(FKN3) $BrO_3^- + Br^- + 2H^+ \rightarrow HBrO_2 + HOBr$
$$\text{Rate} = k_3[BrO_3^-][Br^-][H^+]^2$$

(FKN2) $HBrO_2 + Br^- + H^+ \rightarrow 2HOBr$
$$\text{Rate} = k_2[HBrO_2][Br^-][H^+]$$

Process A $BrO_3^- + 2Br^- + 3H^+ \rightarrow 3HOBr$

These are two electron-transfer processes between the Br(V), Br(III), Br(+I), and Br(−I) oxidation states. Subsequent reactions of HOBr become important in resetting the clock.

When the bromide ion concentration has fallen sufficiently, the major reaction channel for $HBrO_2$ switches from (FKN2) to the reaction with bromate ion, so initiating Process B:

(FKN5) $BrO_3^- + HBrO_2 + H^+ \rightleftharpoons 2BrO_2^{\cdot} + H_2$

Rate = $k_5[BrO_3^-][HBrO_2][H^+] - k_{-5}[BrO_2^{\cdot}]$

(FKN6) $BrO_2^{\cdot} + M_{red} + H^+ \rightarrow HBrO_2 + M_{ox}$

Rate = $k_6[BrO_2^{\cdot}][M_{red}][H^+]$

Process B $BrO_3^- + HBrO_2 + 2M_{red} + 3H^+ \rightarrow 2HBrO_2 + 2M_{ox} + H_2O$

The species $BrO_2^{\cdot}$ involves the Br(IV) oxidation state. Further reaction requires a one-electron transfer step that is provided by the metal ion catalyst.

In the simplest case, for which the reverse step (FKN-5) can be neglected and $[BrO_2^{\cdot}]$ can be expressed as a steady-state balance between (FKN5) and (FKN6)—see Section 1.3—the rate at which there is a net formation of $HBrO_2$ is determined by the rate of step (FKN5). The rate of accumulation thus increases autocatalytically as $[HBrO_2]$ increases. This autocatalytic growth is limited by the self-disproportionation reaction involving $HBrO_2$

(FKN4) $2HBrO_2 \rightarrow BrO_3^- + HOBr + H^+$ Rate = $k_4[HBrO_2]^2$

The switch from Process A to Process B occurs when reactions (FKN2) and (FKN5) are roughly equal. As the bromate concentration remains virtually constant during a given oscillation, the switch to autocatalysis occurs when the $[Br^-]$ has been reduced by Process A to the *critical bromide ion concentration*

$$[Br^-]_{cr} = (k_5/k_2)[BrO_3^-] \approx 1.4 \times 10^{-5}[BrO_3^-] \qquad (3.1)$$

Because bromide ion competes so strongly for $HBrO_2$ it plays the role of an *inhibitor*, delaying the establishment of the autocatalytic feedback via process B.

Resetting the clock

Processes A and B provide for a classic clock reaction. In the induction period beginning at point A in Fig. 3.1, any bromide ion inhibitor (present as impurity or produced in an earlier cycle of the oscillatory reaction) is removed. At the end of the induction period, point B in Fig. 3.1, the bromide ion concentration has fallen to $[Br^-]_{cr}$ as given by Eqn (3.1) and there is then an autocatalytic acceleratory oxidation of the metal ion catalyst via Process B to point C. In the absence of a resetting mechanism, this would be effectively the end of the story. In order to regenerate the starting conditions, a source of bromide ions is needed and the catalyst must be reduced back to its lower oxidation state. These requirements are met simultaneously through Process C, which involves the organic reactant, malonic acid (MA).

A detailed understanding of Process C has only recently begun to emerge. Instead, here we follow the original representation of FKN. This proposes that HOBr can give rise to the bromination of MA, perhaps

L. Györgyi, T. Turanyi, and R.J. Field (1990). *J. Phys. Chem.*, **94**, 7162.

through the formation of Br_2, to produce bromomalonic acid BrMA. Both MA and BrMA react with M_{ox} to yield the reduced form of the catalyst and, in the case of BrMA, bromide ion. Process C is thus represented as

Process C $\quad 2M_{ox} + MA + BrMA \rightarrow fBr^- + 2M_{red} +$ other products
$$\text{Rate} = k_c[\text{Org}][M_{ox}]$$

The stoichiometric factor f provides something on an 'adjustable' parameter. It represents the number of bromide ions produced as two M_{ox} ions are reduced. If M_{ox} reacts solely with BrMA, $f = 2$; for $f > 2/3$ there is a net increase in bromide ion through each oscillatory cycle. In the simplest analysis, f is taken as a constant: in more sophisticated studies, f is allowed to vary with the instantaneous concentrations of MA, BrMA, or of HOBr. A chain mechanism involving malonyl radicals MA· and bromine atom radicals Br· has been proposed to account for stoichiometric factors greater than 2. The rate of Process C, as written above, depends on the total concentration of organic species, [Org], which early on is approximated by the initial concentration of MA.

P. Ruoff, M. Varga, and E. Körös (1988). *Acc. Chem. Res.*, **21**, 326–32.

The rate constants for the steps in the FKN mechanism are given in Table 3.1.

Table 3.1 Rate constants for FKN mechanism and oregonator model

	FKN	Oregonator*
k_1	8×10^9 M^{-2} s^{-1}	6.4×10^9 M^{-1} s^{-1}
k_2	3×10^6 M^{-2} s^{-1}	2.4×10^6 M^{-1} s^{-1}
k_3	2 M^{-3} s^{-1}	1.28 M^{-1} s^{-1}
k_4	3×10^3 M^{-1} s^{-1}	3×10^3 M^{-1} s^{-1}
k_5	42 M^{-2} s^{-1}	33.6 M^{-1} s^{-1}
k_c	1 M^{-1} s^{-1}	1 M^{-1} s^{-1}

*Assumes $[H^+] = 0.8M$

3.3 Conditions for oscillations: Oregonator model

With a reasonable mechanism and the appropriate values for the reaction rate constants, it should be possible not only to match individual experimental observations but also to predict more generally the experimental conditions under which oscillations might be observed. For this, it is especially convenient to use the *Oregonator* model due to Field and Noyes derived from the FKN scheme. This is frequently written in the form

R.J. Field and R.M. Noyes (1974). *J. Chem. Phys.*, **60**, 1877–84.

(O3) $\quad\quad\quad A + Y \rightarrow X + P \quad\quad\quad$ Rate = $k_3 AY$

(O2) $\quad\quad\quad X + Y \rightarrow 2P \quad\quad\quad\quad\;$ Rate = $k_2 XY$

(O5) $\quad$ A + X $\rightarrow$ 2X + 2Z $\quad$ Rate = k_5AX

(O4) $\quad$ 2X $\rightarrow$ A + P $\quad$ Rate = k_4X^2

(OC) $\quad$ B + Z $\rightarrow$ ½fY $\quad$ Rate = k_cBZ

The translation from chemical species to the letters used here is given in Table 3.2. The concentrations of the major reactants, A and B, are treated as constants and [H$^+$] is subsumed into the rate constants.

Table 3.2 Identification of symbols for Oregonator model

A	BrO$_3^-$
B	All oxidizable organic species
P	HOBr
X	HBrO$_2$
Y	Br$^-$
Z	M$_{ox}$

The reaction rate equations for the intermediate species X, Y and Z are

$$dX/dt = k_3AY - k_2XY + k_5AX - 2k_4X^2 \quad (3.2)$$

$$dY/dt = -k_3AY - k_2XY + \tfrac{1}{2}fk_cBZ \quad (3.3)$$

$$dZ/dt = 2k_5AX - k_cBZ \quad (3.4)$$

Dimensionless equations

The next stage is to transform the concentrations X, Y and Z into *dimensionless variables*. This is a step that often stimulates chemists into closing the book, but is really rather painless if handled properly and has definite advantages. The transformation to be used here involves replacing X, Y, Z and T in Eqns (3.2)–(3.4) by x, y, z and τ, defined by

$$x = 2k_4X/k_5A,\ y = k_2Y/k_5A,\ z = k_ck_4BZ/(k_5A)^2 \text{ and } \tau = k_cBt \quad (3.5)$$

Essentially, however, these still represent the concentrations of HBrO$_2$, Br$^-$ and M$_{ox}$ and the time, and can be most usefully thought of in these terms. The advantage of this substitution is that the rate equations now become

$$\frac{dx}{d\tau} = \frac{qy - xy + x(1-x)}{\varepsilon} \quad (3.6)$$

$$\frac{dy}{d\tau} = \frac{-qy - xy + fz}{\varepsilon'} \quad (3.7)$$

$$\frac{dz}{d\tau} = x - z \quad (3.8)$$

Dimensionless parameters

There are three dimensionless parameters:

$$\varepsilon = k_c B/k_5 A, \quad \varepsilon' = 2k_c k_4 B/k_2 k_5 A, \text{ and } \quad q = 2k_3 k_4/k_2 k_5 \qquad (3.9)$$

The first of these depend on the initial concentrations of bromate ion, MA and H^+; q involves only the reaction rate constants. For typical values, $A = 0.06M$ and $B = 0.02M$, we have

$$\varepsilon = 10^{-2}, \quad \varepsilon' = 2.5 \times 10^{-5}, \text{ and } \quad q = 9 \times 10^{-5} \qquad (3.10)$$

The autocatalysis in $HBrO_2$ is revealed by the term $x(1 - x)$ in Eqn (3.6); this is of the same form as the quadratic autocatalysis seen earlier and arises from Process B, step (O5), limited by the disproportionation step (O4). The term $(q - x)y$ in Eqn (3.6) arises from the production and removal of $HBrO_2$ via steps (O3) and (O2), i.e., Process A.

Dynamic steady-state approximation for bromide ion

The other important point involves the size of ε and ε'. Both of these are small, with ε', in particular, being much less than 1. Both of these parameters appear in the *denominator* of a reaction rate equation. Because ε' is so small, the concentration of bromide ion y will change quickly in time (i.e., $dy/d\tau$ will be large) unless the numerator in Eqn (3.7) is also small. This argument is simply a mathematically-based statement of the classic steady-state approximation, so we now assume

$$y = y_{ss} = fz/(q + x) \qquad (3.11)$$

at all times, i.e., the bromide concentration is in a dynamic steady state relative to the $HBrO_2$ concentration. Substituting this result into the reaction rate equations, we have

$$\varepsilon \frac{dx}{d\tau} = x(1 - x) - \frac{(x - q)}{(q + x)} fz \qquad (3.12)$$

$$\frac{dz}{d\tau} = x - z \qquad (3.13)$$

The two terms in Eqn (3.12) represent Process B and Process A respectively, whilst the two terms in Eqn (3.13) described the production of M_{ox} in Process B and its reduction in Process C.

We will not apply the steady-state approximation, based on the magnitude of ε, to $HBrO_2$; partly this is because no simple formula for x_{ss} emerges but also we must retain two concentrations to allow oscillatory behaviour.

Typical behaviour: steady states and oscillations

Equations (3.12) and (3.13) can be integrated numerically for any given choice of f. Figure 3.2(a) shows the resulting variation in x and z for $f = \frac{1}{4}$. After some initial transient development, the concentrations settle to constant, steady-state values. (Steady states different from the chemical

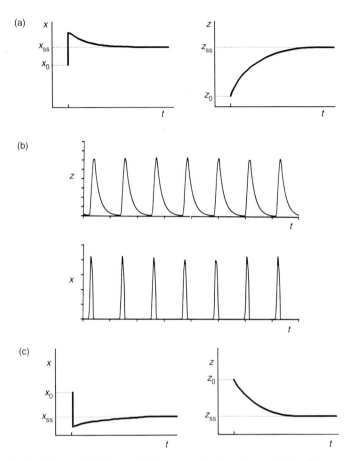

Fig. 3.2 Evolution of HBrO$_2$ and M$_{ox}$ concentrations (x and z) for Oregonator model with (a) $f = \frac{1}{4}$, (b) $f = 1$, and (c) $f = 3$. For (a) and (c) the system evolves from some arbitrary initial condition (x_0, z_0) to a stable steady state (x_{ss}, z_{ss}) of high and low x respectively; for (b) the system evolves to an oscillatory response.

equilibrium state arise in this model because the consumption of the reactants has been neglected.) Under steady-state conditions, the rates of change of both intermediate concentrations become zero simultaneously, $dx/d\tau = dz/d\tau = 0$. Thus $z_{ss} = x_{ss}$, and x_{ss} is given by

$$x_{ss} = z_{ss} = \tfrac{1}{2}\{1 - (f+q) + [(f+q-1)^2 + 4q(1+f)]^{1/2}\} \quad (3.14)$$

The variations of x_{ss}, z_{ss} and of the corresponding steady-state bromide concentration y_{ss} with the stoichiometric factor are shown in Fig. 3.3.

A low value for f corresponds to relatively weak resetting Process C, as few bromide ions are produced as the catalyst is reduced. The corresponding steady-state concentrations of HBrO$_2$ and M$_{ox}$ are thus relatively high (the dimensionless concentrations x_{ss} and z_{ss} are numerically equal, but the actual concentrations $(HBrO_2)_{ss}$ and $[M_{ox}]_{ss}$ will differ due to the different scaling factors involved in the transformations, Eqn (3.5)) whilst the steady state bromide ion concentration y_{ss} is relatively low.

Figure 3.2 (c) shows the evolution of x and z to relatively low steady-state values for the case $f = 3$: the corresponding y_{ss} is relatively high for this strong resetting. For $f = 1$, however, the situation is quite different. A steady-state solution still exists, but the system does not settle to this. Instead a sustained, periodic oscillation in x and z (and hence also in y) about the steady state emerges, Fig. 3.2(b). The steady state is *unstable* for this set of parameter values. The computation may be repeated for various other values of f, with oscillatory behaviour being observed over the range $½ < f < 1 + \sqrt{2}$ indicated by the dashed steady state locus in Fig. 3.3.

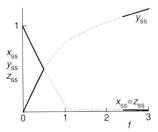

Fig. 3.3 Variation of the steady-state concentrations of $HBrO_2$, Br^-, and M_{ox} (x_{ss}, y_{ss} and z_{ss}) with the stoichiometric factor f. The steady state is unstable for $½ < f < 1 + \sqrt{2}$, over which region the steady-state locus is shown by a broken curve.

3.4 Pictorial representation of oscillations

The reason why the steady state loses its stability over the range of f indicated above can be revealed pictorially. For this, we look not at the evolution of x or z as a function of time, but plot the evolution of one concentration as a function of the other. The two concentrations x and z form a *phase plane*: as x and z vary in time, they draw out a curve or *trajectory* on this plane. We can use the properties of the phase plane to solve Eqns (3.12) and (3.13) graphically and thus obtain the conditions for oscillations without recourse to computation.

There are two special curves that lie on the phase plane, as indicated in Fig. 3.4. These are the *nullclines* that connect x, z pairs for which $dx/d\tau = 0$ (the x-nullcline) or $dz/d\tau = 0$ (the z-nullcline). The z-nullcline is simply the

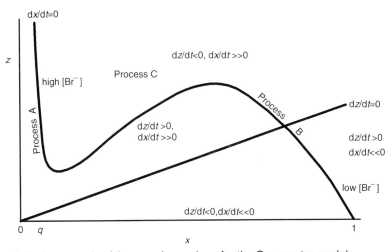

Fig. 3.4 An example of the x–z phase plane for the Oregonator model showing the x- and z-nullclines. The signs of the rate terms for x and z in the four regions are indicated as are the regions in which the three overall processes, A–C, are most important.

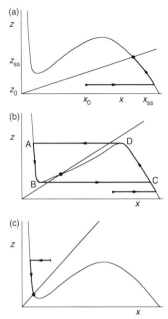

Fig. 3.5 The phase plane description of the evolution of x and z for the three cases shown in figure 3.2: (a) $f = \frac{1}{4}$, (b) $f = 1$, and (c) $f = 3$. In each case, the steady state is indicated by the crossing of the x- and z-nullclines. The system trajectory starts from some arbitrary initial point (x_0, z_0): in (a) and (c), the system eventually approaches the steady state intersection after first having a fast horizontal motion to the appropriate branch of the x-nullcline. For (b), the intersection lies on the middle branch of the x-nullcline and so is unstable. The system cannot approach the intersection along the x-nullcline and instead continuously has fast horizontal jumps separated by slow motions along the left- and right-hand branches of the x-nullcline, so describing a closed loop or 'limit cycle' in the phase plane.

straight line $x = z$ emerging form the origin with unit slope. Any trajectory crossing this line must have a maximum or minimum in z at that point.

The x-nullcline is determined by the more complex condition

$$\frac{(1-x)(q+x)x}{f(x-q)} = z \qquad (3.15)$$

This is a cubic-type curve which asymptotes to $z \to \infty$ as $x \to q$ and has $z = 0$ for $x = 1$; the range of interest is thus $q < x < 1$ and the nullcline has a minimum and a maximum in this range. The two nullclines divide the phase plane into four regions in which the sign of dx/dt and dz/dt vary between positive and negative, indicating the slope of any trajectory in that region. High values for x correspond to low instantaneous concentrations of bromide ion (from Eqn (3.11)) whilst low $HBrO_2$ concentration correspond to high $[Br^-]$. Remembering that Process A corresponds to the reduction of $[Br^-]$, we can recognize that this will occur whilst the system is close to the left-hand branch of the x-nullcline. Along the right-hand branch, Process B is important and z increases (oxidation of the catalyst, increasing z). Process C corresponds to the oxidation of the organic species by M_{ox} and so occurs at high z. This process reduces $[HBrO_2]$ and increases $[Br^-]$, so causing the system to move to the left and down in the phase plane.

The nullclines for $f = \frac{1}{4}$, 1 and 3 are shown in Fig. 3.5. In each case, there is a single point at which the x- and z-nullclines intersect. At such a point, $dx/d\tau = dz/d\tau = 0$, so this locates the steady state solution.

In the case of Fig. 3.5(a), with $f = \frac{1}{4}$, the intersection point lies to the right of the maximum. If we start at some arbitrary initial point (x_0, z_0), corresponding to the initial concentrations of x and z, the subsequent evolution may be predicted by the following argument. If the initial point does not lie on the x-nullcline, the right-hand side of Eqn (3.12) will be nonzero: if the initial point is below the x-nullcline, $dx/d\tau$ will be positive. Also, because of the small parameter ε that effectively occurs as a divisor of the rate expression, $dx/d\tau$ will be large in magnitude. This means that the $HBrO_2$ concentration will change rapidly, on a timescale such that z remains effectively constant. This gives rise to a horizontal movement in Fig. 3.5(a), as indicated, from (x_0, z_0) to some point on the right-hand branch of the x-nullcline. Now $dx/d\tau = 0$. Unless the system has actually jumped to the steady state intersection, $dz/d\tau$ remains nonzero. This gives rise to a slower evolution: if the system lies below the z-nullcline, $dz/d\tau$ will be positive, so the concentration of M_{ox} will increase. As z varies, so x will continually adjust to keep the right-hand side of Eqn (3.12) close to zero. Thus the trajectory undergoes a slow evolution along the x-nullcline until it approaches the steady state intersection. The trajectory illustrated in Fig. 3.5(a) corresponds to the time-series shown in Fig. 3.2(a). A similar argument shows that any initial points finally approach the steady state, although if we start above the x-nullcline, the initial jump is to the left-hand branch of the x-nullcline, followed by a slower evolution down this branch, a second jump from the minimum onto the right-hand branch of the x-nullcline and then the final approach to (x_{ss}, z_{ss}).

With $f = 3$, the intersection lies on the left-hand branch of the x-nullcline, close to the minimum. Again, we can use the argument that any system lying off the x-nullcline will respond by a horizontal jump to one or other of the outer branches of that nullcline followed by a slower evolution along it. In this way, the steady state is finally approached, as indicated by the trajectory corresponding to the time-series in Fig. 3.2(c).

For the case $f = 1$, however, the situation is different. The intersection point now lies on the middle branch of the x-nullcline. If we start at some arbitrary initial point below the x-nullcline, the system jumps to the right-hand branch and then moves upwards towards the maximum (as $dz/d\tau > 0$). At the maximum, $dz/d\tau$ is still positive, so the system must leave the x-nullcline. This gives rise to a horizontal jump to the left-hand branch of the x-nullcline. Now $dz/d\tau < 0$, so the system moves slowly down this branch, towards the minimum. At the minimum, $dz/d\tau$ is still negative, so again the system leaves the x-nullcline. This gives rise to a horizontal jump back to the right-hand branch and the process repeats. At no time can the system jump to the middle branch, and so the steady-state point is never approached. Instead there is a continuous *cycling* in the phase plane around the closed loop ABCD. This loop is called a *limit cycle*: the term 'limit' indicating that this is the ultimate response as $t \to \infty$ and all initial transients have died out. The system evolves to this limit cycle from all initial conditions, i.e., from all starting points in the x–z phase plane, so the amplitude and period of the oscillation are dependent only on the kinetics and the experimental conditions, not the initial condition. Also, if the system receives some later perturbation, it will return to the same limit cycle and hence to oscillations of the same amplitude and period. This indicates that the limit cycle is stable. Stable limit cycles and stable steady states are also termed *attractors* because of this feature of 'attracting' the trajectories in the phase plane.

The points A, B, C, and D correspond to the features of the bromide time-series marked in Fig. 3.1. The maximum bromide ion concentration occurs at point A. The evolution from A to B along the left-hand branch of the x-nullcline corresponds to the gradual reduction of the catalyst, Process C, with the slow blue to red colour change. This is followed by the removal of Br^- through Process A, with $[Br^-]$ reaching the critical bromide ion concentration at point B. This is the condition for the autocatalytic process B to begin, so the $HBrO_2$ concentration increases rapidly, corresponding to the jump BC. The evolution along the branch CD corresponds to the oxidation of the metal catalyst, so z increases, and to the red $\to$ blue colour change. The motion along this branch, although slow compared with the jump in $HBrO_2$, is still relatively fast compared with the motion along AB, so the colour change is observed to be sharp. At point D, the M_{ox} concentration is sufficiently high for Process C to begin in earnest; Br^- builds up in sufficient concentration to inhibit the autocatalysis, so the concentration of $HBrO_2$ falls rapidly and the system returns to the start of the cycle.

3.5 Conditions for oscillations: analysis

The condition for oscillations is simply that the steady state should lie on the middle branch of the x-nullcline, i.e., that x_{ss} lies between the minimum and the maximum. The coordinates of the maximum and minimum in the x-nullcline depend on the parameters q and f, being given approximately by

$$\text{minimum } x = (1 + \sqrt{2})q, \quad z = (1 + \sqrt{2})^2 q/f, \quad y = 1 + 1/\sqrt{2} \quad (3.16)$$

$$\text{maximum } x = \tfrac{1}{2}, \quad z = 1/4f, \quad y = \tfrac{1}{2} \quad (3.17)$$

for $q \ll 1$. The condition for the steady state, Eqn (3.14), to lie at the minimum is then $f = 1 + \sqrt{2}$, whilst for the steady state to coincide with the maximum, we require $f = \tfrac{1}{2}$. The condition for oscillation is thus, as given previously,

$$\tfrac{1}{2} < f < 1 + \sqrt{2} \quad (3.18)$$

Oscillations are suppressed if f, the number of bromide ions produced for every two oxidized catalyst ions reduced in Process C, is either too large or too small. A balance between the efficiency of Process C and the rates of Processes A and B must be achieved. If f is too small, the system settles to a steady state corresponding to the oxidized form of the catalyst and low bromide ion concentration. If f is too large, the build-up of bromide inhibits the autocatalysis and oxidation in Process B and a reduced steady state is established. Qualitative changes in the behaviour of a system as a parameter is varied, such as the change from steady state to oscillatory reaction as f enters the above range, are termed *bifurcations*: the change from steady state to oscillation usually occurs at a *Hopf bifurcation*. Other examples of bifurcations will occur in later chapters.

This analysis relies heavily on the smallness of ε. If ε becomes larger, the jumps in the phase plane will cease to be horizontal as $dx/d\tau$ will be smaller in magnitude and z may vary significantly over the longer timescale for the change in x. As ε increases, the range of f over which oscillations can be observed decreases, with oscillations disappearing completely as $\varepsilon \to 1$. This condition can be interpreted in terms of the reactant concentrations: from (3.9), the requirement $\varepsilon < 1$ becomes

$$[BrO_3^-] > (k_c/k_5)[Org] \approx 0.03[MA] \quad (3.19)$$

The condition on the reactants in terms of f is less easily translated into an explicit form due to the rather mysterious nature of the stoichiometric factor.

3.6 Amplitude and period

A more refined calculation of the critical bromide ion concentration can be obtained from the analysis above. The concentration, y, at the minimum was given above, Eqn (3.21). In dimensional terms, this becomes

$$[Br^-]_{cr} = (1 + 1/\sqrt{2})(k_5/k_2)[BrO_3^-] \approx 2.4 \times 10^{-5}[BrO_3^-] \quad (3.20)$$

c. 70% higher than that for which the rates of steps (FKN2) and (FKN5) actually become equal, indicating that the subsequent autocatalytic growth following step (FKN5) provides Process B with an added advantage in competing with Process A. The experimentally measured value for the ratio $[Br^-]_{cr}/[BrO_3^-]$ is 2×10^{-5}.

The concentrations of the various species at points A and C can also be calculated approximately:

A. $x = q, \quad z = 1/4f, \quad y = 1/8q$ (3.21)

C. $x = 1, \quad z = (1 + \sqrt{2})^2 q/f, \quad y = (1 + \sqrt{2})^2 q$ (3.22)

These can be used to estimate the amplitude and period of the oscillations in terms of the experimental conditions. The Pt electrode responds primarily to the metal ion redox couple: the amplitude in these terms should then be given by

$$A_{Pt} \propto \log([M_{ox,max}]/[M_{ox,min}]) = \log(z_{max}/z_{min}) = -\log[4(1+\sqrt{2})^2 q] \approx 2.7$$

corresponding to a signal of 160mV.

For the bromide electrode,

$$A_{Br} \propto \log([Br^-]_{max}/[Br^-]_{min}) = \log(y_A/y_C) = -\log[8(1+\sqrt{2})^2 q^2] \approx 6.4$$

corresponding to a signal of nearly 400 mV.

To estimate the period of the oscillation we can argue that by far the slowest part of the cycle is the evolution along the branch AB of the x-nullcline. Along this branch, $x \approx q$, so $dz/d\tau = q - z$. Integrating this from z_A to z_B gives

$$\tau_p = -\ln\{4[(1+\sqrt{2})^2 - f]q\} = 6.4 \quad (3.23)$$

for $f = 1$, giving a period of approximately five minutes, slightly longer than observed in practice. We may also note, that the period will vary slowly with the value chosen for f, increasing as f increases.

3.7 Excitability

The BZ system is also exemplary for demonstrating another form of behaviour that will be of particular importance in the next chapter, that of *excitability*. An excitable system is characterized by having a stable steady state, and so is not spontaneously oscillatory. Small perturbation, e.g., by small reductions in the bromide ion concentration, disturbs the system from this state transiently, but the system returns quickly without any change in colour. With slightly larger perturbations, the system is stimulated into a single excursion, with a colour change and back and a single large peak in the intermediate concentrations, similar to a single oscillation, before returning to the original steady state. The system responds in a qualitatively different way to perturbations that are below and above some *critical* or *threshold* value: amplifying those above the threshold.

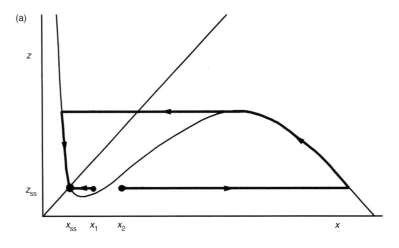

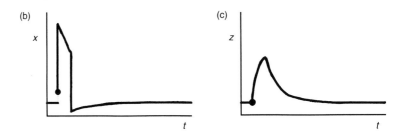

Fig. 3.6 Excitability in the BZ reaction. (a) Phase plane representation: two perturbations of the system from the steady state intersection are shown. The first, small perturbation (x_1) is subcritical: the system returns directly to the steady state with a horizontal fast motion to the left. The second perturbation x_2 is larger and supercritical: the initial fast motion is to the right-hand branch of the x-nullcline along which the system then evolves before a second fast motion back to the left-hand branch, down which there is a final, slow return to the steady state. The corresponding variations of x and z for the excited system are shown in (b) and (c) respectively.

BZ systems which show excitability typically have values for f slightly above the maximum for oscillations. Figure 3.5(c) corresponds to excitable systems with $f = 3$. The phase plane representation of excitability is shown schematically in Fig. 3.6(a). The steady state intersection point lies slightly to the left of the minimum in the x-nullcline. If the concentration of $HBrO_2$, x, is increased by some perturbation but to some value that remains to the left of the middle branch of the x-nullcline, the system simply jumps back to the intersection point. If the perturbation takes the system across the middle branch of the x-nullcline, however, the resulting jump takes the trajectory over to the right-hand branch. The system then must take the 'scenic route' back, moving up the x-nullcline to the maximum, jumping virtually horizontally to the left-hand branch and then moving slowly down that to the steady state. The response in terms of the concentration time series is shown in Fig. 3.6(b).

Two more features are worthy of note. Whilst the system is moving along the right-hand branch of the x-nullcine, it is virtually insensitive to further perturbations, and is termed *refractory*. The system is also effectively refractory as it begins its motion down the left-hand branch, because the middle branch of the nullcline is far away in the phase plane. As the system approaches the steady state, so it regains its excitability. The threshold perturbation decreases, the further down the left-hand branch the system moves and, hence, systems with f only slightly in excess of $1 + \sqrt{2}$ are the most excitable.

3.8 Other oscillatory systems

Oscillations accompany the overall reaction in a number of other chemical processes that can provide the rather moderate demands of feedback and resetting mechanisms. The oldest of these is the Bray–Liebhafsky (BL) reaction, the iodate-catalysed disproportionation of H_2O_2 in acidic solution. This reaction is light-sensitive and not well-characterized mechanistically. Briggs and Rauscher (BR) combined the BL system with the malonic acid resetting of the BZ reaction to produce an oscillator that changes colour from gold to blue to colourless and back to gold. Other solution-phase oscillators include the chlorite–iodide–malonic acid (CIMA) reaction which is related to the bromate–bromide–MA (BZ) reaction by halogen atom substitutions.

W.C. Bray and H.A. Liebhafsky (1931). *J. Am. Chem. Soc.*, **53**, 38.

T.S. Briggs and W.C. Rauscher (1973). *J. Chem. Educ.*, **50**, 496.

P. De Kepper, J. Boissonade, and I.R. Epstein, (1990). *J. Phys. Chem.*, **94**, 4404–12.

In the gas-phase, the oxidation of carbon monoxide may proceed in an oscillatory manner. The oscillations are to all intents and purposes isothermal. Typically, less than ½ % of the fuel is consumed in each oscillation. The oscillatory evolution gives way to steady reaction well before complete consumption, with typically 50% of the fuel remaining at the end of the oscillatory phase. The oscillations seem strongly dependent on the surface-coating or age of the reactor surface, implying some heterogeneous contribution to the mechanism: this reaction will be discussed further in Chapter 6. *Cool flames* accompanying the oxidation of simple hydrocarbon fuels, and especially their partially-oxygenated derivative such as acetaldehyde (ethanal, CH_3CHO) also occur widely and are related to important phenomena such as engine knock. Again, these will be discussed in Chapter 6.

S.K. Scott (1991). *Chemical chaos*, Oxford University Press.

J.F. Griffiths (1986). *Adv. Chem. Phys.*, **64**, 203–303.

Oscillations can also be observed in the decomposition of formic acid by sulphuric acid (the Morgan reaction). This is an example of a *gas evolution oscillator* (GEO). The product, CO, has a limited solubility in the aqueous solvent, but the solution can become supersaturated to a relatively high degree. Eventually, some chance nucleation leads to a dissolution event that returns the CO_{aq} concentration to its equilibrium value. Further production via the decomposition reaction leads to a further supersaturation, followed by a later dissolution event. The kinetics of nucleation are virtually discon-

R.M. Noyes (1990). *J. Phys. Chem.*, **94**, 4404–12.

tinuous, providing a strong nonlinear feedback route. The decomposition of aqueous ammonium nitrite provides another example of a GEO.

An important class of oscillators involves the transport of a key reactant from one phase to another. The best studied example is the partial oxidation of benzaldehyde to benzoic acid, catalysed by cobalt (Co^{2+}/Co^{3+}) and Br^- ions, which draws on the atmosphere above the solution surface for O_2. The related partial oxidation of *p*-xylene to terephthalic acid is a multimillion tonne per annum process to provide one of the monomers for polyethylene terephthalic acid (PET) and is also oscillatory.

In biological systems, periodicity is also common. Enzymatic catalysis provides for nonlinearity via the relatively familiar Michaelis–Menten rate law. Feedback may also arise quite commonly with *allosteric* enzymes: these consist of more than one subunit, each with its own active site where binding and subsequent reaction can occur. The binding of a substrate at one site may affect the activity of the remaining subunits: with positive feedback or *activation*, the activity of other sites is enhanced; with negative feedback or *inhibition*, the activity is reduced. The change in activity is effected by changes in the spatial conformation of the enzyme between *relaxed* and *tensed* forms. Allosteric effects underlie the oscillations observed, both *in vivo* and *in vitro* during glycolysis.

B. Hess and A. Boiteux (1971). *Ann Rev. Biochem.*, **40**, 237–58; (1980) *Ber. Bunsenges. Phys. Chem.*, **84**, 346–51.

Another biochemical system exhibiting oscillatory behaviour is the peroxidase–oxidase reaction in which organic electron donors are oxidized by molecular oxygen. An important electron donor is nicotinamide adenine dinucleotide hydride NADH which is also involved in the storage and recovery of energy in biological systems. This is typically not run as a closed system, but is provided with a continuous inflow (but not outflow) of reagents. Such semi-batch reactors have some advantages over closed systems, but also share some of the disadvantages that we can now discuss.

L.F. Olsen and H. Degn (1977) *Nature*, **267**, 177–8.

3.9 Reactant consumption and limitations of closed systems

Closed systems are relatively easily set up, transported and operated. Oscillations in cooperative reactions, such as the BZ reaction, may last for a long while—but they cannot last forever. During the course of each oscillation in the intermediate concentrations, small but nonzero amounts of the reactants are consumed, so providing the overall driving force for the reaction. Strictly speaking, therefore, each oscillation occurs against a slightly different background concentration of reactants: each oscillation is inevitably slightly different from its predecessor and from subsequent excursions. Closed systems are good for lecture demonstrations but bad for serious research. Sooner or later, the reactant depletion will become sufficiently great so as to take the mixture out of the composition range in which oscillations occur, and the system will settle to the more familiar, quasi-steady evolution for its final approach to the chemical equilibrium state.

4 Targets, spirals, and scrolls

Reactions that show oscillations in closed systems give rise to a range of interesting wave behaviour in unstirred systems. As described in Chapter 2, clock reactions give rise to one-off reaction-diffusion fronts, with the reactants being completely consumed as the front passes. In BZ-type systems, the front corresponds to a wave of reaction involving the intermediates, with little consumption of the major reactants. The resetting mechanism means that there will be a wave 'back' following the front, returning the system to its original, prefront state. This gives rise to a *pulse*. If there are further initiations, the system will also be able to support further waves, possibly giving rise to a series of pulses, a *wavetrain*. These wavetrains can organize the reaction mixture in two or three dimensions into a variety of intricate, evolving patterns.

4.1 Kinematic and phase waves

Two of the types of wave observed in so-called *active media*, such as the BZ reaction, have little to do with diffusion. If the BZ reagent is distributed in a tube along which a spatial gradient of say pH or temperature is established, the period of the spontaneous oscillation will vary continuously in space. This can give rise to an apparent travelling wave, as the red → blue transition occurs at slightly different times at each position. This *kinematic wave* is of the same character as the 'moving' lights of a neon advertising sign. A *phase wave* also has little involvement of diffusion, relying on the reaction being initiated at different times as we move along the tube. There is no background spatial gradient, but the spontaneous oscillation again occurs at a time that varies smoothly along the tube, giving the appearance of movement. The speeds at which such waves 'travel' is determined primarily by the underlying concentration of phase gradients and so can be tuned to almost any value. Phase waves do play a role in creating the patterns in the BZ and other systems, but only when coupled with true reaction-diffusion or *trigger* waves.

J. Ross, S.C. Müller, and C. Vidal (1988). *Science*, 240, 460.

4.2 Trigger waves in one dimension: pulses

The BZ system supports pulses or wavetrains for compositions such that the bulk reaction is spontaneously oscillatory. The situation is, however, somewhat simpler if the conditions are changed slightly, so that the reaction mixture is spontaneously excitable (see Section 3.7). This means that the reaction sits at steady state in the absence of perturbations. Small perturba-

R.J. Field and R.M. Noyes (1974). *J. Am. Chem. Soc.*, **96**, 2001. See also J.J. Tyson (1976). The Belousov–Zhabotinskii reaction, in *Lecture notes in biomathematics*, **10**, Springer-Verlag, Berlin; and (1985) in *Oscillations and travelling waves in chemical systems*, (ed. R.J. Field and M. Burger), Chap. 3, p. 93, Wiley, New York.

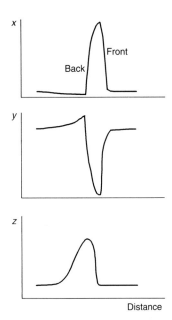

Fig. 4.1 Schematic representation of $HBrO_2$, Br^-, and M_{ox} profiles for a single trigger wave pulse propagating into an excitable BZ mixture showing an excitation front followed after some delay by a recovery wave. The wave is propagating from left to right.

tions lead only to a rapid return to the steady state but disturbances above some critical threshold cause a single, large excursion before returning.

An excitable system subjected to a single, sufficiently large initiation will support a single wave pulse. The front of this pulse is very little different from the fronts described in Chapter 2. Ahead of the front, the mixture is in the reduced state (coloured red). The front is a wave of oxidation, that causes an increase in $HBrO_2$ and M_{ox}, with a consequent decrease in Br^-, as sketched in Fig. 4.1. Immediately after the wave, there is a period during which $[HBrO_2]$ falls only slowly. This quasi-steady region is the equivalent of the post-front composition in the iodate–arsenite system. In the BZ system, however, there is a wave back following some distance behind. This is a reduction wave, through which $HBrO_2$ and M_{ox} are consumed and Br^- is produced, returning to their pre-front concentrations.

The wave propagates at a constant velocity and is driven by the diffusion of $HBrO_2$ into the reduced state ahead of the front. There, it reduces the local Br^- concentration (via Process A reactions). Once $[Br^-]$ decreases below $[Br^-]_{cr}$, the autocatalytic Process B leads to the local oxidation and production of $HBrO_2$ as in the homogeneous oscillation. The autocatalysis in the BZ system gives rise to a quadratic-type rate law, $x(1-x)$, as seen in Eqn (3.11), for example. The analysis in Section 2.2, Eqn (2.13), showed that the speed of such reaction-diffusion fronts is given by

$$c = 2(Dk)^{1/2} \tag{4.1}$$

where D is the diffusion coefficient of the autocatalytic species and k is the pseudo-first-order rate constant for the autocatalytic reaction. The autocatalysis in the BZ system involves $HBrO_2$ and is provided by

$$\text{Process B} \quad BrO_3^- + HBrO_2 + 2M_{red} + 3H^+ \rightarrow 2HBrO_2 + 2M_{ox} + H_2O \tag{4.2}$$

This has a rate determined by

$$\text{Rate} = k_5[BrO_3^-][H^+][HBrO_2] \tag{4.3}$$

The appropriate form for k in Eqn (4.1) is thus

$$k = k_5[BrO_3^-][H^+] \tag{4.4}$$

so

$$c = 2(Dk_5[BrO_3^-][H^+])^{1/2} \tag{4.5}$$

The prediction that the speed of the pulse front varies with $[BrO_3^-]^{1/2}$ and $[H^+]^{1/2}$ has been confirmed experimentally. For typical values, $D = 2 \times 10^{-5}$ cm^2 s^{-1}, $k_5 = 42$ M^{-2} s^{-1}, $[BrO_3^-] = 0.06$M and $[H^+] = 0.8$M, this predicts a wave-speed of 2.4 mm min^{-1}, which is typical of the magnitudes observed.

4.3 Waves in two dimensions: circles and targets

Many of the studies of waves in the BZ system have been carried out in thin films of solution in a Petri dish. Again, the simplest situation arises when the reaction mixture is excitable, rather than spontaneously oscillatory. A point source initiation then naturally gives rise to an expanding circular wave pulse. Once these have expanded sufficiently, they propagate much as one-dimensional waves. In the presence of a periodic or continuous stimulation, a series of concentric rings, giving rise to a *target patterns* of blue circles nested on a red background, emerges, as shown in Fig. 4.2. The inner rings rely on the resetting of the intermediate concentrations and so are not possible in one-off clock-type reactions.

The spontaneous initiation of the waves in a target pattern appears to rely on the presence of heterogeneities, such as the presence of dust particles or defects in the surface of the Petri dish. Careful filtering suppresses this spontaneous initiation in excitable systems. Each target reflects the frequency of the initiation at its centre. If this is caused by random heterogeneities, different target cores may naturally have different frequencies. The speed of the waves varies from one target pattern to another, with low frequencies leading to higher velocities. These two effects mean that the waves are more closely spaced for high frequency sources, i.e., these have a shorter wavelength.

When waves from two targets collide, they annihilate each other. As the subsequent wave for the higher frequency pattern comes closer on the heels of the annihilated wave than that for the low frequency source, the next collision occurs closer to the latter. Over the course of several collisions, waves from the high frequency source eventually reach the vicinity of the low frequency core, which is then also annihilated. High frequency targets grow at the expense of low frequency ones; given sufficient time, the highest frequency source will entrain the whole dish.

For a given target pattern, Fig. 4.2 also reveals that the outermost wave travels at a slightly higher speed than subsequent, inner waves, all of which have essentially the same speed. This indicates a dependence of the velocity on the concentrations ahead of the wave that does not emerge in the simple analysis leading to Eqn (4.5). This point is addressed below.

Recipe due to A.T. Winfree for BZ target patterns: mix 67 ml H_2O, 2 ml H_2SO_4 and 5 g $NaBrO_3$. Take 6 ml and add ½ ml of a solution of 1 g NaBr in 10 ml H_2O and 1 ml of a solution of 1 g malonic acid in 10 ml H_2O. Wait for the Br_2 produced to vanish. Add 1 ml of 2mM ferroin and wait for waves to develop. Other recipes are collated by W. Jahnke and A.T. Winfree (1991). *J. Chem. Educ.*, **68**, 320.

4.4 Modelling wavetrains

The two variable Oregonator model of Section 3.3, and the pictorial method of Section 3.4, is of great value for the description of the wavetrains in one dimension and the related target patterns in two dimensions. Equations (3.12) and (3.13) must be augmented to allow for diffusion, giving

$$\varepsilon \frac{dx}{d\tau} = \varepsilon^2 \nabla^2 x + x(1-x) - \frac{(x-q)}{(q+x)} fz \qquad (4.6)$$

$$\frac{dz}{d\tau} = \varepsilon \nabla^2 z + x - z \qquad (4.7)$$

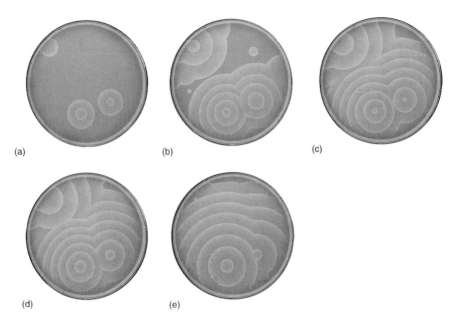

Fig. 4.2 Target patterns in a thin layer of BZ reaction in a 9 cm diameter petri dish. (a) 1 min after mixing; (b) after 3 min 30s; (c) after 7 min 15s; (d) after 7 min 35s; (e) after 16 min 20s. Three random pacemaker sites initiate targets, but as the system evolves, the successive annihilations of colliding waves from two adjacent targets occur closer and closer to the lower frequency pacemaker. In time, the higher frequency source entrains the lower one. In a given target, the outermost wave travels at a slightly higher velocity than those inside the target. (Photographs courtesy of M. Pearson.)

These equations, including the Fick's law diffusion terms $\nabla^2 x$ and $\nabla^2 z$, are dimensionless, with x and z representing [HBrO$_2$] and [M$_{ox}$] respectively. The appearance of the terms ε^2 and ε in front of the diffusion terms indicates that the diffusion coefficients of HBrO$_2$ and M$_{ox}$ are relatively small quantities compared with the size of the Petri dish and the kinetic timescale. For a one-dimensional system, $\nabla^2 x = \partial^2 x/dr^2$, where r is the dimensionless distance: for two dimensions, the Laplacian can be written in circular polar coordinates and the circular symmetry then gives $\nabla^2 x = \partial^2 x/dr^2 + (1/r)\partial x/\partial r$, where r is now the dimensionless radial distance from the centre of the target.

Again, we can use the x–z phase plane and the nullclines of the kinetic terms, to represent the wave pictorially. With an excitable system, the intersection point of the nullclines occurs to the left of the minimum in the x-nullcline as indicated in Fig. 4.3. The steady-state composition corresponds the reaction mixture sufficiently far ahead of any advancing wave front. As a front approaches, the diffusion of HBrO$_2$ increases x locally. If x is increased beyond the threshold value, corresponding to the middle of the x-nullcline, then the local kinetics will cause the system to jump across the

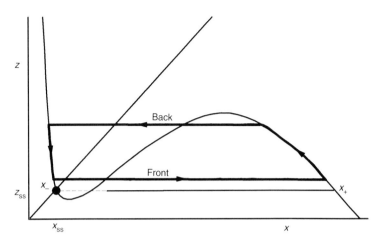

Fig. 4.3 Phase plane representation of BZ waves. Initially the system sits at the steady state indicated by the intersection of the x- and z-nullclines. An increase in [HBrO$_2$] (x) may arise due to diffusion from an approaching front. Once this perturbs the system sufficiently far from the steady state (across the middle branch of the x-nullcline as indicated by dashed line) the system responds with a fast motion across the phase plane to the right-hand branch of the nullcline: the concentration of the oxidized form of the catalyst remains virtually constant during this jump, with $z = z_{ss}$. This corresponds to the outermost wavefront of a target pattern. The system then undergoes slow motion along the right-hand branch of the x-nullcline, during which oxidation of the metal catalyst and hence a colour change occurs, until it approaches the maximum. Close to the maximum, there is a sharp transition back to the left-hand branch, corresponding to the wave back. In this recovery process, the system then evolves along the left-hand branch, with z decreasing. At some stage a further front is initiated, but this usually occurs before z regains its steady state concentration and so lies at higher z.

phase plane to the right-hand branch. This jump can be identified with the leading front of the wave pulse.

The jump occurs whilst the concentration of the oxidized catalyst z remains effectively constant and equal to its steady-state value. Rather than dealing with two reaction-diffusion equations, therefore, we need only consider the modified form of Eqn (4.6)

$$\varepsilon \frac{dx}{d\tau} = \varepsilon^2 \nabla^2 x + x(1-x) - \frac{(x-q)}{(q+x)} f z_0 \qquad (4.8)$$

where z_0 is a constant, and $z_0 = z_{ss}$ in this case. Apart from the final term, Eqn (4.8) is essentially the same as Eqn (2.8), with a quadratic autocatalytic rate law rather than a cubic form. In fact, as $x_{ss} \approx q$, the second term makes little contribution at the leading edge of the front, which is where the wave-speed is mainly determined. As the system departs from the left-hand branch and x increases, however, the final term in Eqn (4.8) makes its presence felt. The value of x ahead of the front x_- is simply the steady-state value, $x_- = x_{ss}$: after the jump, $x \approx 1$, with the exact value x_+ given by the highest solution of the nullcline condition

$$x(1-x)(q+x)/(x-q) = fz_0 \qquad (4.9)$$

for the same $z_0 = z_{ss}$. The wave-speed is determined by z_0 and also by the concentrations x_- and x_+ ahead of and behind the front.

In the post-front region, the system, moves along the right-hand branch of the x-nullcline, just as observed for the well-stirred system. The concentration of the oxidized catalyst, z, therefore also changes in this region. At some point, there is a jump back to the left-hand branch corresponding to the wave back. This jump may occur when the system reaches the maximum in the nullcline, or it may happen earlier, stimulated by diffusional loss of x. If the pulse is to maintain a constant shape, the speed of the wave back must equal that of the front. The wave-speed will again be determined by Eqn (4.8), but this has now to be solved for some higher value of z_0; the initial and final values for x, x_+, and x_-, are also different from those appropriate to the front. If a back with the appropriate speed does not exist, this *recovery* wave is simply a phase wave.

Once the system has returned locally to the left-hand branch of the x-nullcline, it slowly evolves along this towards the steady state. For a one-off circular wave pulse, the steady state will finally be attained (strictly speaking at infinite time). For target patterns, however, the system may not have time to return to the vicinity of the steady-state point. At some time determined by the frequency of the source, a subsequent front will approach. Provided the system has recovered sufficiently to have left its refractory phase and regained its excitability, the diffusion of $HBrO_2$ from this new wave will increase x beyond its (slightly higher) threshold and stimulate a new jump to the right. This occurs for some z_0 that is higher than z_{ss} and so the wave will have a different (lower) speed than the first for which $z_0 = z_{ss}$. All subsequent waves will jump from the same z_0, whose value will depend on the source frequency, and so have the same speed. If we concentrate on the behaviour at any given point in the reaction medium, locally the solution is simply undergoing a regular oscillation similar to that described in the previous chapter.

Targets with different frequency sources will have different effective values of z_0 and hence different speeds, as observed experimentally. The dependence of the wave-speed on the source frequency or, equivalently, the period or wavelength of the target pattern is known as the *dispersion relation*. This can be determined (numerically) from Eqn (4.8) which gives the speed, and Eqn (3.23) which gives the period.

J.J. Tyson and J.P. Keener (1988). *Physica D*, **32**, 327.

4.5 Curvature effects

The above analysis is appropriate to plane waves or to circular waves that have grown sufficiently large in radius to have low curvature. For the initial development of circular waves at the centre of a target, however, there is generally high curvature. Typical sizes of the source for target patterns indicate a diameter of the order of 90 μm: for a circle, the curvature is simply the inverse of the radius, $\kappa = -1/r$. With high curvatures, the dilution due to diffusional spreading of the autocatalyst into the region ahead of the wave is much enhanced compared to the planar wave case. If this dilution is too strong, the wave may fail as the perturbation will not reach the threshold required to reduce [Br⁻] to its critical value. The relationship between the speed of a planar wave, c_0, and that of an identical wave with curvature c, is given by the *eikonal equation*

$$c = c_0 - D/r \tag{4.10}$$

where D is the diffusion coefficient. If the curvature becomes too high, i.e., if r is too low, the speed of the wave may become zero, indicating failure. This *critical radius for initiation* is given by setting $c = 0$ in Eqn (4.10), giving

$$r_{cr} = D/c_0 \tag{4.11}$$

Using the result $c_0 = 2.4$ mm min^{-1} ($= 4 \times 10^{-3}$ cm s^{-1}) and taking $D = 2 \times 10^{-5}$ cm^2 s^{-1}, we have $r_{cr} = 50$ μm, in good agreement with the observed value. We may also note that where waves collide in Fig. 4.2, to create cusp-shaped regions with positive curvature, the wave-speed in enhanced relative to the planar wave.

4.6 Spirals

If target waves are broken, e.g., by gently tilting the reaction mixture, the two ends created respond by curling up to form a pair of counter-rotating spirals. Spiral waves thus formed are shown in Fig. 4.4. In contrast to the behaviour of targets, all spirals in a given BZ reaction mixture rotate with the same period and wavelength, reflecting the bulk kinetics rather than local heterogeneities. (The outermost turn of the spiral propagates at a slightly higher speed than the inner turns as it is propagating into reagent at the steady-state composition; the inner turns are propagating into the wake of an earlier wave, exactly as for target waves.) The propagation speed must be constant at all places in the bulk, or the spiral will deform as it evolves. This conservation of shape provides a second condition that selects just one of the wave speed-period pairs from the dispersion curve available to target patterns. Spirals with multiple arms have also been created experimentally; Fig. 4.5 shows examples with 1, 2, 3, and 4 arms rotating about a common core.

A.T. Winfree (1972). *Science*, **175**, 634: (1974). *Sci. Amer.*, **240**, 82.

J.P. Keener and J.J. Tyson (1986). *Physica D*, **21**, 307.

The advent of modern imaging systems coupled to computer acquisition hardware has led to an increase in the detail with which spirals have been studied. At the centre of a given spiral is a region of the order of 350 μm in which the concentration variations are distinctly less than in the development front. At the centre of this region is a *core* with a diameter of the order of 30 μm. In the simplest form of spiral evolution, the tip of the spiral, where the oxidation front and the reduction back meet, rotates around this circular core with the same frequency as the bulk spiral, as indicated in

S.C. Müller, T. Plesser and B. Hess (1985). *Science*, **230**, 661.
A. Pagola and C. Vidal (1987). *J. Phys. Chem.*, **91**, 501.

48 *Targets, spirals, and scrolls*

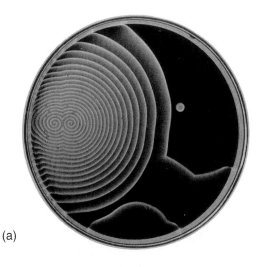

(a)

Fig. 4.4 Spiral waves in the BZ reaction formed by breaking target patterns. Each spiral rotates with the same period (approximately 30s) and has the same frequency, although the outermost wavesegment of each spiral propagates with a slightly higher velocity than the inner turns. (Photographs courtesy of M. Pearson.)

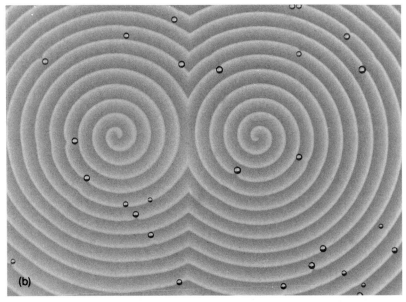

(b)

A.T. Winfree (1991). *Chaos*, **1**, 303; W. Jahnke and A.T. Winfree (1991). *Int. J. Bifurcation and Chaos*, **1**, 445.

Fig. 4.6(a). As the conditions are varied, however, the tip of the spiral begins to *meander*. Initially, this involves a second oscillation appearing, so the tip describes a quasiperiodic motion. As the behaviour becomes more complex, various patterns resembling flowers or other objects are described by the tip path as shown in Fig. 4.6(c). This change from 'rigid' to 'compound' rotation occurs as the excitability of the system is decreased. This can be achieved experimentally by decreasing the concentration of bromate or H$^+$. In terms of the dimensionless parameters in the Oregonator model, this corresponds to increasing the value of f, so that the intersection point moves further up the left-hand branch of the x-nullcline away from the minimum, or of increasing the value of ε, so that the two species HBrO$_2$ and M$_{ox}$ (x and z) evolve on less different timescales. The petal patterns are essentially

Oscillations, Waves and Chaos 49

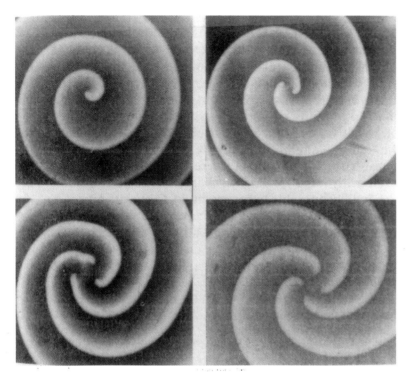

Fig. 4.5 Multi-armed spirals in the BZ reaction, showing 1–, 2–, 3–, and 4– rotors surrounding a single core. (Reprinted with permission from K.I. Agladze and V.I. Krinsky (1982). *Nature*, **296**, 424: © Macmillan Journals, London.)

'phase-locked' quasiperiodic motion, with the two oscillatory frequencies having some simple ratio, so that the actual path closes up. Further complexities arise as the excitability is decreased further and the size of the petal shape shows a general increase as f increases. Eventually, as the system becomes sufficiently non-excitable, the system may return to simple quasi-periodic meandering, then to rigid rotation and, finally, spiral wave behaviour can no longer be supported.

Spiral waves are also observed in the growth cycle of the slime mould *Dictyostelium discoideum*. A colony of this beast will feed in a given area until the nutrient falls below some critical concentration. Certain cells then begin to emit periodically a messenger chemical cyclic adenosine monophosphate (cAMP). The surrounding cells then migrate towards these pacemakers, proceeding up the concentration gradient (a feedback process known as *chemotaxis*). This gives rise to spirals in the cell density that are virtually indistinguishable from the BZ spirals in black and white photographs.

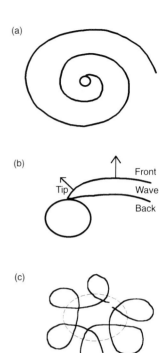

Fig. 4.6 The tip of a spiral does not lie at a fixed point in space. (a) In the simplest form of motion, rigid rotation, the tip rotates around a core (the rotation is anticlockwise in this case: see enlargement). (b) At lower excitability, the path followed by the tip shows a meandering or compound rotation that creates petal-shaped flower patterns.

A.J. Durston (1973). *J. Theor. Biol.*, **42**, 483.

4.7 Waves in three dimensions: scroll waves

A single spiral can be imagined lying in an extremely thin reaction layer, but in thicker layers or other genuine three-dimensional situations, other shapes

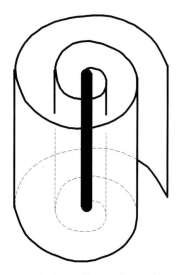

Fig. 4.7 A scroll wave formed by stacking identical spiral with constant phase. The stack of cores forms a filament which, in this case is straight and untwisted.

J.J. Tyson and J.P. Keener (1991). *Physica D*, **53**, 151.

J.D. Murray (1990). *Mathematical biology*, Springer-Verlag, Berlin. L. Glass and M.C. Mackey (1988). *From clocks to chaos; the rhythms of life*, Princeton University Press, New Jersey.

A.M. Turing (1952). *Phil. Trans. R. Soc. Lond.*, **B237**, 37.

are possible. The simplest situation would arise if the solution effectively gives rise to a set of spiral waves stacked up exactly on top of each other. The resulting wave would then resemble a scroll of paper unwinding, as sketched in Fig. 4.7 (or, at least, a folded sheet of paper unwinding: the waves have fronts and backs and a tip corresponding to the fold). The tips of these stacked spirals would then revolve around a stack of cores, describing a thin circular tube or *filament*. In this simplest case, the filament is straight. In other situations, the filament may bend or twist. This arises because the unwinding spiral sheets distort and stretch. In some cases, the inner sheets of the spirals may collide with outer turns, leading to annihilations. Some very complex three dimensional patterns are created with the filament forming a loop or becoming knotted, giving rise to highly dramatic three-dimensional wave structures.

4.8 Significance in biology

Spiral and scroll waves occur in some important areas outside chemistry. Waves of *spreading depression* have been observed associated with the depolarization of the neuronal membrane of the brain, spiralling around lesions in the brain cortex. The correlation between such wave activity and such abnormalities as epilepsy is under current study. Scroll waves have been linked reasonably firmly with *cardiac arrhythmias*. Under normal circumstances, heart muscle produces a coordinated contraction with a frequency of $c.$ 1 Hz; in *ventricular flutter*, larger amplitude contractions occur with a frequency of $c.$ 7 Hz and are associated with a rotating spiral wave in the heart tissue; in *fibrillation*, the contractions are smaller in amplitude and occur in an apparently random manner, with potentially fatal consequences. The onset of fibrillation seems to be connected with the development of scroll-type waves in the muscular wall of the heart. Elsewhere, calcium waves pass over the surface of a fertilized egg as a precursor to cell division. The enhancement in velocity achieved by coupling diffusion to a feedback process is exploited for nerve signal transmission: repetitive signalling requires a resetting mechanism. The Hodgkin–Huxley model for this system essentially involves a voltage and an ion current (due to Na^+/K^+ and Ca^{2+}/Mg^{2+} balances across the nerve membrane) which evolve in a phase plane containing a cubic-type nullcline intersected by a linear nullcline that is very similar to that for the BZ system.

4.9 Patterns: diffusion-driven instabilities

In 1952, Alan Turing predicted theoretically that chemical systems showing feedback in situations where not all species diffuse with the same mobility might give rise to *spontaneous pattern formation*. An initially well-mixed, homogeneous solution would, then, evolve of its own volition to produce spatial gradients in the participating species, with local enhancements or depletions in the concentrations of some of the species. If such *diffusion-driven instabilities* arise in the BZ or iodate–arsenite systems, they would

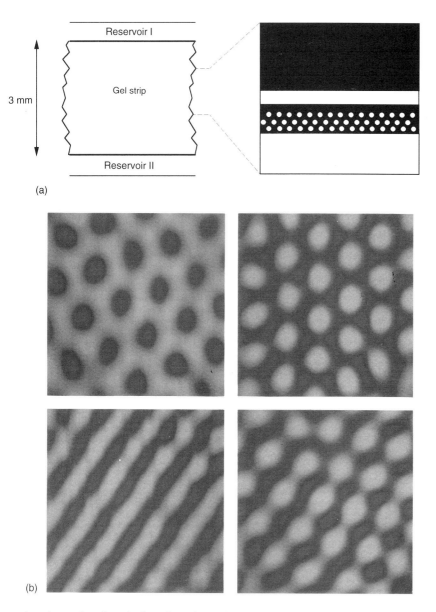

Fig. 4.8 Diffusion-driven (Turing) and other patterns in chemical systems. (a) Turing patterns in the chlorite–iodide–malonic acid reaction in a gel-strip reactor: continuous flows of reactants maintain constant concentrations at the top and bottom edges of the 3 mm strip and establish concentration gradients in that direction. The Turing 'spots', however, arise perpendicular to these gradients indicating their diffusion-kinetic origin. Dark and light regions correspond to high and low $[I_3^-]/[I_2]$ (complexed with thiodene indicator) respectively. (Reprinted with permission from V. Castets, E. Dulos, J. Boissonade and P. De Kepper (1990). *Phys. Rev. Lett.*, **64**, 2953. © American Physical Society.) (b) Spotted and striped Turing patterns for the same reaction in a thin circular gel. Constant concentrations (different for each response) are maintained on front and back surfaces, but there are no imposed concentration gradients in the plane of the pattern. (Reprinted with permission from Q. Ouyang and H.L. Swinney (1991). *Chaos*, **1**, 411. © American Institute of Physics.)

give rise to local variations in colour that would reveal the *Turing patterns*. Turing suggested that this kinetic-diffusion process might underlie the spontaneous development of spatial form in morphogenesis, e.g., of a developing embryo. The mechanism has also been implicated in the spots and stripes marking animal coats and how these change with the size of the animal.

The main condition on the diffusion coefficients is that for the autocatalytic species must be sufficiently less than that for the reactant. For the iodate–arsenite system this would require $D_{I^-} < D_{IO_3^-}$. Under normal circumstances, ions have diffusion coefficients that do not vary much from one species to another, so Turing patterns have not been widely observed in chemical systems. Very recently, however, the first examples have been obtained by carrying out a variation of the iodate–arsenite reaction in a gel into which a complexing agent that binds I^- or I_3^- has been incorporated (Fig. 4.8). Whilst not perhaps quite as dramatic as spots on a leopard's coat, this experimental confirmation of the basic mechanism proposed over 40 years ago represents an important advance.

5 Bistability and oscillations in flow reactors

The disadvantages of studying reactions in thermodynamically closed vessels, in which anything other than the equilibrium state is strictly transient, have been remarked upon at various stages in the previous chapters. The use of open reactors that have a continuous inflow of fresh reactants and a matching outflow of products overcomes many of these, and the analysis of the behaviour is also made easier if the reactor is well stirred so there are no spatial gradients. Well-stirred flow reactors (CSTRs or continuous-flow well-stirred tank reactors) are also widely exploited in practical situations in the chemical industry, and their study—chemical reaction engineering—forms an important part of chemical engineering courses.

Because of the continuous flow, molecules spend only a finite time in the reactor: the average time spent is called the *mean residence time*, t_{res}. This means that there is not infinite time for the reaction to approach the chemical equilibrium state appropriate to the inflow concentrations. The system is constrained 'far from equilibrium'. In the simplest cases, the reaction proceeds to a true *steady state*, at which the net rate of inflow of each reactant species exactly balances its net rate of chemical removal by reaction, and the rate of chemical formation of each product or intermediate species balances its rate of outflow. There are, however, other possibilities, especially when the reaction allows for feedback.

5.1 Steady states and bistability

The iodate–arsenite and iodate–bisulphite reactions have been studied in flow reactors. We may anticipate the observed behaviour to some extent. If the reactants are pumped through the reactor at very high flow rates, the correspondingly very short residence times will be insufficient for much reaction to occur. The steady-state concentrations established will not differ much from the inflow composition. On the other hand, with very low flow rates, the residence time becomes very long and the system has almost enough time to approach the chemical equilibrium composition.

Figure 5.1(a) shows the variation of steady-state iodate and iodide concentrations established within a CSTR as a function of the flow rate. At low flows, the system does indeed approach the equilibrium state (virtually complete conversion of iodate to iodide); this segment of the curve is known as the *thermodynamic branch* as it approaches the thermodynamic equilibrium. At high flow rate, the concentrations approach their inflow values

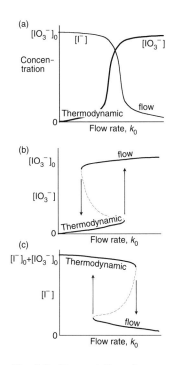

Fig. 5.1 The variation of steady-state iodate and iodide ion concentrations with flow rate for the Landolt reaction in a CSTR: (a) thermodynamic and flow branches merge smoothly so there is a unique steady state for each flow rate; (b), (c) thermodynamic and flow branches overlap to provide a region of bistability and the associated phenomenon of hysteresis.

$[IO_3^-]_0$ and $[I^-]_0$ respectively along the *flow branch* of the steady-status locus. For the system shown in Fig. 5.1(a), the thermodynamic and flow branches connect through a continuous variation in the steady state concentrations. At any given flow rate there is just one steady state composition: the steady state concentration of iodate increases and that of iodide decreases as the flow rate increases (or, equivalently, as the residence time decreases). Although the curve is continuous, there is a region of relatively rapid variation, similar to the sharp change in the time-dependent concentrations observed in the clock reaction involving these species.

If the operating conditions are varied somewhat, an important change may occur to the steady state locus. Figures 5.1(b,c) show the variation of the steady-state concentrations in which the locus has become 'folded' so that the thermodynamic and flow branches overlap, rather than connect. For a range of flow rates, the system has more than one steady state to choose from. We will see that only the uppermost and lowermost are accessible, so this phenomenon is known as *bistability*. If the reaction is started with a sufficiently high flow rate, beyond the region of bistability, the system will settle onto the unique steady state lying on the flow branch. If the flow rate is decreased in small increments, the system will essentially move along the flow branch as the steady-state concentrations vary slightly; it will remain on the flow branch as the region of bistability is traversed. As the flow rate is decreased further, however, the flow branch turns back on itself—the system must now 'fall off' this branch and move to the thermodynamic branch. There is thus an abrupt change in the steady-state composition: the steady-state concentrations actually change discontinuously with the flow rate at such fold points: discontinuous responses to continuous changes in the operating conditions are known as *bifurcations* and this is an example of a *saddle-node* bifurcation, for reasons explained later. Figures 5.1(a–c) are examples of *bifurcation diagrams* in which the response of the system is plotted as a function of the experimental conditions.

If the flow rate is now increased again, the system will remain on the thermodynamic branch through the region of bistability, until the second fold point is encountered. The jump back to the flow branch thus occurs at a higher flow rate than the jump away from it. There is *hysteresis* as the system is cycled around the S-shaped steady-state curve by these changes in flow rate. Within the region of bistability, the actual steady state selected depends not only on the flow rate but also on the previous history of operations within the reactor.

G.A. Papsin, A. Hanna, and K. Showalter (1981). *J. Phys. Chem.*, **85**, 2575; P. De Kepper, I.R. Epstein, and K. Kustin (1981). *J. Am. Chem. Soc.*, **103**, 6121.

5.2 Mass balance equations for flow reactors

We may use the simple model fo the iodate–arsenite reaction, involving just cubic autocatalysis, to illustrate the form and analysis of the equations governing the rates of change of the different species concentrations within a well-stirred flow reactor.

In the presence of excess reductant, the iodine concentration remains very low and can be treated as if in dynamic steady state with the concentrations of iodate and iodide. The reaction can be represented by the single stoichiometric process

$$IO_3^- + 5I^- + 6H^+ \to 6I^- + 3H_2O$$
$$\text{Rate} = R_\alpha = (k_{\alpha 1} + k_{\alpha 2}[I^-])[I^-][IO_3^-][H^+]^2 \quad (5.1)$$

In a closed system, taking $k_{\alpha 1} \approx 0$,

$$-d[IO_3^-]/dt = d[I^-]/dt \approx k_c[IO_3^-][I^-]^2 \quad (5.2)$$

where $k_c = k_{\alpha 2}[H^+]^2$ so that the reaction is assumed to be buffered. This has the same form as the cubic autocatalysis rate law (1.21) with $A = IO_3^-$ and $B = I^-$ as remarked earlier. In a closed system, we also have the conservation condition on iodine atoms:

$$[IO_3^-] + [I^-] = [IO_3^-]_0 + [I^-]_0 \quad (5.3)$$

at all times, where $[IO_3^-]_0$ and $[I^-]_0$ are the initial concentrations. Using the symbols a and b, the rate equation can then be written in any of the following ways:

$$-da/dt = db/dt = k_c ab^2 = k_c a(a_0 + b_0 - a)^2 = k_c(a_0 + b_0 - b)b^2 \quad (5.4)$$

In a flow reactor, there is an additional process that acts with the kinetics to change the species concentrations, namely, the flow process itself. The total rate at which the concentrations of iodate and iodide are changing is related to the instantaneous concentrations by the *mass balance equations*

$$V d[IO_3^-]/dt = q\{[IO_3^-]_0 - [IO_3^-]\} - VR_\alpha \quad (5.5)$$

$$V d[I^-]/dt = q\{[I^-]_0 - [I^-]\} + VR_\alpha \quad (5.6)$$

| Rate of change | Inflow – outflow | Reaction rate |

Here, V is the reactor volume and q is the total volumetric flow rate through the reactor. If there is a single inflow, containing all the reactants, then the inflow concentrations $[IO_3^-]_0$ and $[I^-]_0$ are simply the concentrations of the reactants in that stream. If, however, there are separate inflows for each reactant (to prevent reaction being initiated in the inflow) then these inflow concentrations must allow for the dilution that occurs when the inflows are mixed: if there are two inflows with equal flow rates, then $[IO_3^-]_0$ and $[I^-]_0$ will be one half of the actual concentrations of the stock solutions being pumped to the reactor.

Each term in the mass balance equations has units of mol s^{-1}. If we divide through by the volume, the quotient q/V arises naturally. This has units of s^{-1} and is the *flow rate*, k_0. Its inverse is the mean residence time, t_{res}, of the molecules in the reactor.

The conservation of iodine atoms, Eqn (5.3), also still applies with the initial concentrations replace by the inflow concentrations. Thus, using the a,b notation we can rewrite Eqn (5.5) as

$$da/dt = k_0(a_0 - a) - k_c ab^2 = k_0(a_0 - a) - k_c a(a_0 + b_0 - a)^2 \quad (5.7)$$

The second form does not involve b: because we have a single stoichiometric process, the evolution of the system within the reactor is fully described by a single mass balance equation.

5.3 Steady state solutions: flow diagrams

Equation (5.7) is a cubic equation that could, with some effort, be integrated analytically. This, however, is unnecessary: the concentration of iodate will evolve from any given initial value to a steady state that causes da/dt to become equal to zero. Thus, we seek the steady states that satisfy

$$k_0(a_0 - a_{ss}) = k_c a_{ss}(a_0 + b_0 - a_{ss})^2 \tag{5.8}$$

The steady state will depend on the four parameters, k_0, a_0, and b_0, and k_c, which actually also involves the H$^+$ concentration. However, we know that a_{ss} must lie in the range $0 \leq a_{ss} < a_0$ on simple physical grounds. It helps to rewrite the steady-state condition in dimensionless terms:

$$\kappa_0(1 - \alpha_{ss}) = \alpha_{ss}(1 + \beta_0 - \alpha_{ss})^2 \tag{5.9}$$
$$\quad\quad\quad \mathcal{L} \quad\quad\quad\quad \mathcal{R}$$

Here $\alpha = a/a_0$, so $0 \leq \alpha \leq 1$, is the concentration of iodate relative to its inflow concentration. This form of the equation has only two parameters: $\beta_0 = b_0/a_0$, the relative inflow concentrations of autocatalyst I$^-$ and reactant IO$_3^-$ and $\kappa_0 = k_0/k_c a_0^2$ which is the ratio of the flow rate to a pseudo-first-order rate constant for the autocatalytic step. It is these ratios that determine the qualitative form of the steady-state locus rather than the absolute values of the individual components.

Equations (5.8) or (5.9) are cubics in terms of the steady-state concentrations, but can be solved relatively easily 'the other way round' for a given choice of β_0, substituting in different values of a_{ss} or α_{ss} and calculating the corresponding value for κ_0. Another approach, which also illustrates how the value of β_0 influences the steady-state locus, is to use a *flow diagram*, similar to the thermal diagram introduced in Section 2.1. The reaction term $\mathcal{R}$ on the right-hand side of Eqn (5.9) describes a cubic curve if plotted against $1 - \alpha_{ss}$, as shown in Fig. 5.2(a), which is simply a variation on curve (b) of Figure 1.1. There is a nonzero intersection point on the $\mathcal{R}$-axis, $\mathcal{R} = \beta_0^2$, that depends on the ratio of inflow concentrations. The curve in Fig. 5.2(a) corresponds to a system with a relatively high inflow concentration of the autocatalytic species I$^-$, so $\beta = 0.2$. The flow line $\mathcal{L}$ emerges from the origin and has a slope determined by the dimensionless flow rate κ_0. The intersection of $\mathcal{R}$ and $\mathcal{L}$ indicates the steady-state point. As the flow rate is varied, with κ_0 increasing, so the flow line becomes steeper and the intersection point moves to the left, corresponding to the lower extents of conversion of iodate to iodide: α_{ss} increases so the dimensionless steady-state concentration of iodide, given by

$$\beta_{ss} = 1 + \beta_0 - \alpha_{ss} \tag{5.10}$$

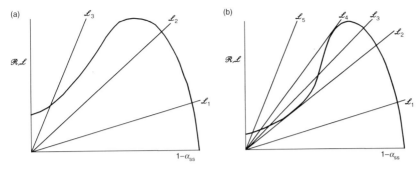

Fig. 5.2 Flow diagram representation of steady states. (a) $\beta_0 > \frac{1}{8}$, three different flow lines $\mathscr{L}_{1-3}$ are shown, corresponding to three different flow rates $\kappa_{0,1-3}$, with $\kappa_{0,1} < \kappa_{0,2} < \kappa_{0,3}$. In each case, $\mathscr{R}$ and $\mathscr{L}$ intersect only once. (b) $\beta_0 < \frac{1}{8}$. Now $\mathscr{R}$ and $\mathscr{L}$ intersect three times for some range of flow rates, $\mathscr{L}_2 < \mathscr{L} < \mathscr{L}_4$. For $\mathscr{L} = \mathscr{L}_{2,4}$ the flow line touches the reaction rate curve tangentially.

decreases. There is always a single intersection point, indicating a unique steady state. The resulting steady-state locus has the qualitative form of Fig. 5.1(a).

Figure 5.2(b) shows the flow diagram for a lower inflow concentration of autocatalyst relative to reactant, $\beta_0 = 0.05$. The reaction curve $\mathscr{R}$ intersects the rate axis much closer to the origin. Five different flow lines, $\mathscr{L}_{1-5}$ are shown, each corresponding to a different dimensionless flow rate $\kappa_{0,1-5}$. For the lowest flow rate, $\kappa_{0,1}$, corresponding to the lowest gradient, the flow line intersects the rate curve at a high extent of reaction, corresponding to low α_{ss} and high β_{ss}. At the highest flow rate, $\kappa_{0,5}$, there is again a single intersection, now lying close to the origin, so α_{ss} is close to 1 and β_{ss} close to β_0. In between these extremes, e.g., for the flow rate $\kappa_{0,3}$, $\mathscr{R}$, and $\mathscr{L}$ intersect three times, allowing for multiplicity of steady states. This gives rise to the region of bistability of the form shown in Fig. 5.1(b,c). The limits of the bistable region, corresponding to the turning points in the steady-state locus, arise when $\mathscr{R}$ and $\mathscr{L}$ meet tangentially, as indicated by $\mathscr{L}_2$ and $\mathscr{L}_4$ with flow rates $\kappa_{0,2}$ and $\kappa_{0,4}$ respectively.

5.4 Turning points and tangencies

The condition for a turning point in the steady-state locus in this model is easily determined using the tangency condition: we require $\mathscr{R} = \mathscr{L}$ and $d\mathscr{R}/d\alpha_{ss} = d\mathscr{L}/d\alpha_{ss}$, i.e.,

$$\kappa_0(1 - \alpha_{ss}) = \alpha_{ss}(1 + \beta_0 - \alpha_{ss})^2 \tag{5.11}$$

$$-\kappa_0 = (1 + \beta_0 - \alpha_{ss})(1 + \beta_0 - 3\alpha_{ss}) \tag{5.12}$$

Dividing one equation by the other, we obtain a quadratic for the steady-state concentration α_{ss} at the turning points, which has roots

$$\alpha_{ss}^{\pm} = \tfrac{1}{4}\{3 \pm (1 - 8\beta_0)^{1/2}\} \tag{5.13}$$

58 *Bistability and oscillations in flow reactors*

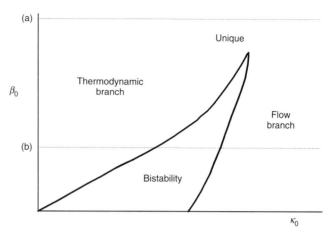

Fig. 5.3 Region of bistability in the β_0–κ_0 (i.e., inflow concentration of I⁻ verses flow rate) parameter plane for cubic autocatalytic model for Landolt reaction. The region of bistability is bounded by the conditions for tangency in the flow diagram: the two tangents merge and bistability is lost for $\beta_0 = 1/8$.

with the corresponding *critical* flow rates $\kappa_0^{\pm}$ given by

$$\kappa_0^{\pm} = -(1 + \beta_0 - \alpha_{ss}^{\pm})(1 + \beta_0 - 3\alpha_{ss}^{\pm}) \tag{5.14}$$

Equation (5.13) only has real roots provides $\beta_0 \leq 1/8$, i.e., if

$$[\mathrm{I}^-]_0 \leq 1/8\, [\mathrm{IO}_3^-]_0 \tag{5.15}$$

The variation of $\kappa_0^{\pm}$ with β_0 is shown in Fig. 5.3. There are two curves that meet at a cusp. Within the region bounded by the curves, the reaction system shows a multiple steady-state solutions (bistability) and hysteresis: outside there is just a single steady state. As β_0 is increased, the region of bistability decreases, and the steady-state locus unfolds completely for $\beta_0 > 1/8$. This unfolding is also sometimes called *transition*. Equation (5.15) thus gives a necessary condition for multiplicity in this system: if the flow provides too high a concentration of iodide, the feedback provided by the autocatalytic kinetics becomes swamped.

5.5 Nodes and saddles: stability of coexisting steady states

The middle branch of solutions, shown as a dashed curve in Fig. 5.1(b,c), has not entered into our discussion much so far. The steady state of the system switched between the thermodynamic and flow branches as the flow rate was varied, but the system never moved onto the middle branch. In fact, although the compositions represented by this section of the bistability loop satisfy the steady-state condition, they are *unstable* states and cannot be realized in practice under normal operating conditions. The existence of these unstable, *saddle point* solutions is, however, significant and they do play an important role in helping decide which of the two stable state branches the system actually selects.

If we choose a flow rate from the region of multiple steady states and denote the three steady states as α_1, α_3 in order of increasing α (iodate ion), then we can identify as corresponding *potential V*,

$$V = \int \{\alpha(1 + \beta_0 - \alpha)^2 - \kappa_0(1 - \alpha)\} d\alpha$$
$$= \tfrac{1}{2}(1 + \beta_0)^2\alpha^2 - \tfrac{2}{3}(1 + \beta_0)\alpha^3 + \tfrac{1}{4}\alpha^4 - \kappa_0\alpha(1 - \tfrac{1}{2}\alpha) + V_0 \quad (5.16)$$

where V_0 is some arbitrary constant related to our choice of energy zero. If V is plotted as a function of α, as shown in Fig. 5.4, the extrema correspond to the steady-state solutions, with α_1 and α_3 corresponding to minima and α_2 to the maximum. This allows a simple mechanical analogy. The steady state on the thermodynamic and flow branches correspond to potential wells in which the system can lie—they are stable solutions. The middle solution is unstable in the same sense as a pendulum balanced upright or a pencil balanced on its end: in theory this state could be achieved from very special initial conditions, i.e., if we start the reaction with α and β exactly equal to α_2 and the corresponding β_2. However, even if this is arranged, the slightest perturbation will send the system off to one or other of the stable states lying in the wells. This gives the middle solution the character of a saddle point. If we start with $\alpha < \alpha_2$, the system will evolve to α_1: if we start with $\alpha > \alpha_2$, it goes to α_3; the saddle point thus *separates the basins of attraction* of the coexisting stable steady states (the coexisting *attractors*).

If the conditions are varied, e.g., by altering the flow rate, so that the boundary of the region of multistability is approached then one minimum and the maximum will merge as these two steady states vanish. Steady states corresponding to potential wells are known as *nodes*, so this coalescence is termed a *saddle-node bifurcation*.

5.6 Designing oscillatory reactions from bistable systems

The nonlinear feedback process that support multiple steady states also frequently can provide the support for oscillatory behaviour. In flow systems, the resetting of the clock discussed in Chapter 3 can be achieved more simply by the continuous direct inflow of the exhausted reactants at a flow rate appropriately matched to the chemical rates. The iodate–arsenite system described above cannot itself become oscillatory, as the reaction is determined by a single stoichiometric process and so has only one *degree of freedom*: once the concentration of iodate, say, has been specified, all other concentrations are fixed by the stoichiometry and the inflow concentrations. In order to have oscillations, two degrees of freedom are necessary: this allows a two-dimensional phase plane in which a limit cycle can lie, whereas a one-degree-of-freedom system has only a phase 'line' onto which a limit cycle cannot be placed.

The close connection between bistability and oscillations has been extensively exploited to design a large family of relaxation oscillators in flow systems, so much so that they constitute by far the largest class of chemical

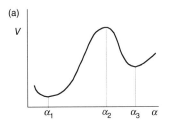

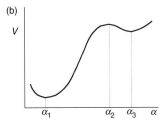

Fig. 5.4 Variation of potential with concentration of iodate ion: (a) the three extrema correspond to the steady states ($dV/d\alpha = 0$), with the minima $\alpha_{1,3}$ corresponding to stable states and the maximum α_2 corresponding to an unstable state determining the regions of attraction of α_1 and α_3; (b) as in (a), but now the parameters have been changed so that α_2 and α_3 are about to merge at the end of the region of bistability, leaving only the flow branch solution α_1.

P. De Kepper and J. Boissonade (1985) in *Oscillations and travelling waves in chemical systems*, (ed. R.J. Field and M.J. Burger), Chap. 7, p. 223. Wiley, New York.

oscillators. The basic principles and experimental method underlying the design of such systems can again be illustrated with the cubic autocatalysis representation of the Landolt iodate–reductant reaction. Some approximations will ease the algebra without misrepresenting the method. The important concepts can be represented pictorially.

Model of a chemical oscillator

The autocatalytic reaction is appended with an equilibrium process involving the autocatalyst and a second species C

$$A + 2B \rightarrow 3B \qquad \text{Rate} = k_c ab^2 \qquad (5.17)$$

$$B + C \rightleftharpoons BC \qquad \text{Rate} = k_1 bc - k_{-1} x \qquad (5.18)$$

where $x = [BC]$. We will assume that C and BC do not flow into or out of the reactor and that the initial concentrations of these species are c_0 and 0 respectively. At any time then $c = c_0 - x$. We also assume $a_0 + b_0 = a + b$, i.e., we expect the amount of B bound up in the BC complex to be small, so $a = a_0 + b_0 - b$.

If $\gamma = x/a_0$ and $\gamma_0 = c_0/a_0$, the mass balance equations can be written much as before, except that it is more convenient to work with β rather than α:

$$d\beta/d\tau = \kappa_0(\beta - \beta) + \beta^2(1 + \beta_0 - \beta) - \kappa_1\beta(\gamma_0 - \gamma) + \kappa_{-1}\gamma \qquad (5.19)$$

$$d\gamma/d\kappa = \kappa_1\beta(\gamma_0 - \gamma) - \kappa_{-1}\gamma \qquad (5.20)$$

Typically κ_1 and κ_{-1} might be small quantities indicating that the equilibrium is established on a much slower timescale than the rate of change in the autocatalyst concentration.

In the absence of the equilibrium process, the dimensionless mass balance equation for the autocatalyst, equivalent to Eqn (5.7) would have the form

$$d\beta/d\tau = \kappa_0\beta_0 \quad - \quad \kappa_0\beta \quad + \quad \beta^2(1 + \beta_0 - \beta) \qquad (5.21)$$
$$\text{Inflow} \qquad \text{Outflow} \qquad \text{Reaction}$$

In the present case, Eqn (5.19) can be rearranged slightly to give

$$d\beta/d\tau = (\kappa_0\beta_0 + \kappa_{-1}\gamma) - [\kappa_0 + \kappa_1(\gamma_0 - \gamma)]\beta + \beta^2(1 + \beta_0 - \beta) \qquad (5.22)$$
$$\text{'Inflow'} \qquad \text{'Outflow'} \qquad \text{'Reaction'}$$

The second reaction step effectively modifies the form of the inflow and outflow terms. First, we may note that the effective inflow and outflow rates are now 'decoupled': in Eqn (5.21) the effective flow rate κ_0 is the same for both flow terms, but in Eqn (5.22) there are different additional terms for the inflow and outflow. Second, the effective inflow and outflow rates involve the concentration γ and so are determined by the instantaneous state of the system. Rather than remaining a fixed constant during the reaction, the flow terms involve a dynamic response variable, with

$$\kappa_{0,\text{eff}} = \kappa_0\beta_0 + \kappa_{-1}\gamma \qquad (5.23)$$

for the inflow.

Pictorial representation

For suitable values of β_0 and κ_0, Eqn (5.21) gives rise to multiple steady states, as described earlier. If we first imagine that γ is a parameter we can control, Eqn (5.22) will also allow multiple steady-state solutions for the autocatalyst concentration over some range of the effective flow rate $\kappa_{0,\text{eff}}$, as shown in Fig. 5.5. When we then acknowledge that the x-axis really involves one of the response variables, we see that Fig. 5.5 really has more in common with the phase plane representations used in Chapter 3 than with the steady state bifurcation diagrams above. If the reactor has a high concentration of B at any moment, the concentration of the complex BC will increase, so the effective inflow rate increases via its definition in Eqn (5.23). This would cause the system to move 'to the right' along the upper branch in Fig. 5.5, with β also falling as BC is produced. If the system reaches the turning point, there will be a jump to the lower branch. Now β is small, so the equilibrium will shift to the left, reducing γ and hence causing $\kappa_{0,\text{eff}}$ to decrease. The system will move to the left along the lower branch, with β increasing as B is formed via dissociation of BC. If this 'trajectory' reaches the left-hand turning point, there would be a jump back to the upper branch—and the cycling would continue to provide sustained oscillations: from bistability to oscillation.

To complete the analysis, we must add a second curve to the phase plane representation. The folded curve seen already is actually the β-nullcline, along which $d\beta/d\tau = 0$. The γ-nullcline is defined by

$$\gamma = \kappa_1 \gamma_0 \beta / (\kappa_{-1} + \kappa_1 \beta) \qquad (5.24)$$

The position of this nullcline relative to the β-nullcline depends on the total concentration of the complexing agent γ_0 and on the equilibrium constant $K_{eq} = \kappa_1/\kappa_{-1}$. There are four possibilities, as indicated in Fig. 5.6(a–d). In Fig. 5.6(a), the two nullclines cross three times, indicating that the system has three steady states, $\beta_1 < \beta_2 < \beta_3$. In Fig. 5.6(b), the relative positions have shifted, so there is now only one intersection. This occurs for high β and so similar to β_3. Like that steady state, this is a stable solution, so the system is attracted to this point and then remains there.

In Fig. 5.6(c), there is again only one intersection, giving rise to a steady state with low β. This is similar to β_1 and is also stable, so again the system is attracted to the steady state. The final situation, represented in Fig. 5.6(d) has the intersection occurring on the middle branch of the fold. The steady

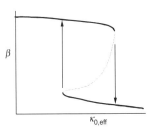

Fig. 5.5 Variation of the pseudo-steady state for β with the effective inflow rate $\kappa_{0,\text{eff}}$ showing a folded locus similar to the bistability curve in the absence of C. If the system reaches one of the turning points, there will be a fast motion vertically, due to the smallness of κ_1 and κ_{-1}, as indicated by the arrows.

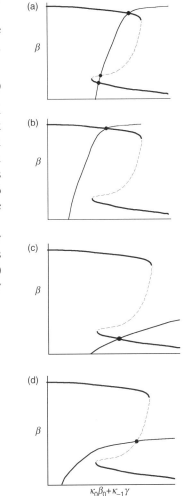

Fig. 5.6 As in Fig. 5.5, but now also showing the γ-nullcline. (a) Three intersections corresponding to a system with three steady states: this system will exhibit bistability, with the intersections on the uppermost and lowest branches of the β-nullcline being stable. (b) One intersection, corresponding to a unique stable steady state with high autocatalyst concentration. (c) One intersection, corresponding to a unique stable steady state with low autocatalyst concentration. (d) One intersection, corresponding to a unique steady state. This lies on the middle branch of the β-nullcline and so is unstable. This system will exhibit sustained oscillations of the form indicated in Fig. 5.5.

state is similar in concentration to β_2, and like that solution is typically unstable. The system will, in this case, describe a limit cycle formed from the upper and lower branches of the fold with vertical jumps at the turning points.

Cross-shaped diagrams

The changes in the multiplicity and stability of the steady-state solutions just described can be brought about by suitably varying the actual flow rate κ_0 and, say, the concentration γ_0, or their dimensional equivalents k_0 and c_0. Figure 5.7 shows the regions in which the different types of response are found. As c_0 increases, the region of bistability (a) narrows as the locations of the two turning points move closer together. These meet at a cusp point, much as shown in Fig. 5.3. However, in this case, the curve seems to continue, giving rise to a *cross-shaped diagram*. Beyond the crossing point lies region (d), corresponding to Fig. 5.5(d) and hence to conditions for which oscillations may be found. Outside the regions of bistability or oscillations are the regions for stable-steady states of the types in Figs 5.5(b) and (c) as indicated.

Figure 5.6 is typical of the diagrams observed experimentally as two of the control parameters are varied. Usually, the region of oscillations closes up again at high c_0 and beyond this is a region in which a unique steady state with autocatalyst concentration similar to β_2 is found. In these cases, β-nullcline will have unfolded in the phase plane. Close inspection reveals that the boundary of the oscillatory region is not simply an extension of the boundaries of the region for bistability. Rather these are separate curves, that lie very close near to the cusp of the bistability region. It is, in fact, possible to have multiple steady states and oscillations, but typically only over narrow ranges of parameter combinations. In these cases, either the uppermost or the lowest steady state loses its stability, so the system moves away if given a small perturbation, whilst the other remains stable. The system now chooses between the stable steady state or a stable limit cycle that emerges around the unstable steady state. If the system settles onto the limit cycle, small amplitude oscillations about the unstable steady state are observed. The unstable saddle point (middle solution) also continues to exist and separates those initial conditions that evolve to the two different coexisting attractors.

For some reactions, and typically over only very narrow ranges of the parameters, both the uppermost and lowest steady states can become unstable. Each may then be surrounded by a separate limit cycle, corresponding to two different oscillatory responses for the same experimental conditions. This bistability of oscillatory solution is known as *birhythmicity*.

I.R. Epstein, K. Kustin, P. De Kepper, and M. Orban (1983). *Sci. Am.*, **248**, 96.

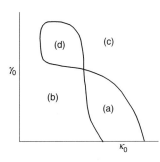

Fig. 5.7 A cross-shaped diagram showing the regions of different behaviour in the γ_0–κ_0 parameter plane. The regions (a)–(d) correspond to the different phase plane portraits in Fig. 5.6, so region (a) corresponds to bistability. As the original system is perturbed by the inflow of species C, so bistability gives way to oscillations, region (d).

5.7 Applications of cross-shaped diagram technique

Various bistable chemical reactions have been converted into relaxation oscillators via the above approach. The search for a chemical oscillator then begins with finding a system that shows bistability, such as the iodate–arsenite system. The next step is to find an additional reactant that somehow provides an additional feedback process. The earliest example of this method involved the iodate–arsenite system to which chlorite ion was added: ClO_2^- plays a slightly different chemical role than the simple complexing equilibrium indicated above, but the basic principle remains intact. The idea of using chlorite ion arose as it is known to react with iodide, the product, to produce iodate, the reactant, and hence is a candidate for chemically 'resetting' the Landolt clock. This feedback pathway

$$3ClO_2^- + 2I^- \rightarrow 2IO_3^- + 3Cl^- \tag{5.25}$$

is catalysed by I_2, a crucial intermediate in the Landolt reaction.

De Kepper, I.R. Epstein, and K. Kustin (1981). *J. Am. Chem. Soc.*, **103**, 2133. This proved the value of the basic method, even though the chlorite–iodide subsystem was later shown to be an oscillatory reaction in its own right.

The iodate–bisulphite reaction has also been used to produce an oscillatory Landolt reaction by introducing an additional inflow of ferrocyanide ion to the reactor. Ferrocyanide is capable of acting in its own right as the reductant in the iodate clock, so this provides something of a mixed clock reaction. An example of the experimental cross-shape diagram for this system is shown in Fig. 5.8

Under the conditions for which the oscillatory Landolt (or EOE) system was studied, the main feedback processes appear to involve pH changes as sulphite is oxidized to the stronger acid sulphate ion. A relatively simple two variable model which accounts for the observed oscillatory waveform and the conditions under which they are observed can be written as

E.C. Edblom, M. Orban, and I.R. Epstein (1986). *J. Am. Chem. Soc.*, **108**, 2826; E.C. Edblom, L. Györgyi, M. Orban, and I.R. Epstein (1987). *J. Am. Chem. Soc.*, **109**, 4876; V. Gaspar and K. Showalter (1987). *J. Am. Chem. Soc.*, **109**, 4869; V. Gaspar and K. Showalter (1990). *J. Phys. Chem.*, **94**, 4973.

(1) $$SO_3^{2-} + H^+ \underset{k_{-1}}{\overset{k_1}{\rightleftharpoons}} HSO_3^-$$

(2) $$HSO_3^- \overset{k_2}{\rightarrow} H^+$$

(3) $$2H^+ \overset{k_3}{\rightarrow} I_2$$

(4) $$I_2 + HSO_3^- \overset{k_4}{\rightarrow} 3H^+$$

(5) $$I_2 \overset{k_5}{\rightarrow} products$$

The mass balance equations for $[HSO_3^-]$ and $[H^+]$ are then

$$[HSO_3^-]/dt = k_1[SO_3^{2-}]_{ss}[H^+] - (k_{-1} + k_2 + k_4[I_2]_{ss} + k_0)[HSO_3^-] \tag{5.26}$$

$$d[H^+]/dt = -k_1[SO_3^{2-}]_{ss}[H^+] + (k_{-1} + k_2 + 3k_4[I_2]_{ss})[HSO_3^-] - 2k_3[H^+]^2 - k_0([H^+]_0 - [H^+])$$

with

$$[SO_3^{2-}]_{ss} = (k_{-1}[HSO_3^-] + k_0[HSO_3^-]_0)/(k_1[H^+] + k_0) \tag{5.27}$$

and $$[I_2]_{ss} = k_3[H^+]^2/(k_4[HSO_3^-] + k_5 + k_0) \tag{5.28}$$

The concentrations of sulphate ion and iodine are taken to be in dynamic steady state with the main variables, bisulphite and hydrogen ions. The Dushman reaction involving iodate and iodide is assumed to be fast, leading to the apparent transmutations of elements. Typical values for the rate

G. Rabai, M. Orban, and I.R. Epstein (1990). *Acc. Chem. Res.*, **23**, 258.

Table 5.1 Typical values for rate constants in model of Oscillatory Landolt (EOE) reaction

k_1	5×10^{10} M^{-1} s^{-1}
k_{-1}	8.1×10^3 s^{-1}
k_2	0.06 s^{-1}
k_3	7.5×10^4 M^{-1} s^{-1}
k_4	2.3×10^9 M^{-1} s^{-1}
k_5	30 s^{-1}
k_0	1.5×10^{-3} s^{-1}
$[HSO_3^-]_0$	0.09 M

constants are given in Table 5.1, with oscillations being predicted over a range of the inflow concentration $[H^+]_0$.

The EOE reaction is one of a number of *pH oscillators* that have been discovered. Some other oscillators designed by this method are listed in Table 5.2. Various additional complexities in the behaviour of such redox reactions have also been observed. Some reactions exhibit *tristability*, with up to three coexisting stable steady states (and two coexisting unstable saddle states). The iodate–arsenite reaction shows *mushroom* and *isola* steady state loci if the reactor is fed with additional stream of solvent. These patterns are illustrated in Fig. 5.9. The mushroom curve shows two separate regions of bistability: the new region, at low flow rates, allows the iodide concentration to be *washed-out*. As the neck of the mushroom skrinks, so the upper branch and the saddle branch become disconnected from the low $[I^-]$ steady state. The system will not jump spontaneously to the high $[I^-]$ branch no matter how the flow rate is varied, even though these solutions are stable. The system must be perturbed sufficiently, e.g., by the addition of iodide ion, to reach the upper branch, and it will fall off this if the flow rate is either made too high or too low.

Table 5.2 Epsteins's taxonomy of inorganic chemical oscillators showing links between different 'families' of components. (Reprinted from I.R. Epstein (1987). *Chemical & Engineering News*, **65**, (March) p.30. © American Chemical Society.)

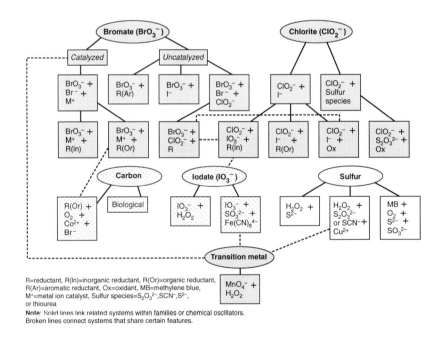

Mushroom and isola patterns can also be found with the cubic autocatalysis model

$$A + 2B \to 3B \qquad \text{Rate} = k_c ab^2 \qquad (5.29)$$

if coupled with a simple, first-order termination step for the autocatalyst

$$B \to C \qquad \text{Rate} = k_t b \qquad (5.30)$$

The flow diagram technique can be applied to this model, with the flow line $\mathcal{L}$ now depending in a slightly more complex manner on the flow rate k_0 (and involving the termination rate constant k_t). If the CSTR is fed by streams of both A and B, with $b_0/a_0^2 < \frac{1}{8}$ and if k_t is sufficiently small (typically $k_t/k_c a_0^2 < \frac{1}{16}$) then up to four tangencies of $\mathcal{R}$ and $\mathcal{L}$ become possible as k_0 is varied.

The mass conservation condition now becomes $a + b + c = a_0 + b_0$, so we cannot use $b = a_0 + b_0 - a$ as before in Eqn (5.7). The mass balance equations for a and b become

$$da/dt = k_0(a_0 - a) - k_c ab^2 \qquad (5.31)$$

$$db/dt = k_0(b_0 - b) + k_c ab^2 - k_t b \qquad (5.32)$$

For a steady state, $da/dt = db/dt = 0$. Adding the steady state equations yields

$$k_0(a_0 + b_0 - a_{ss}) = (k_0 + k_t) b_{ss} \qquad (5.33)$$

We can use this to obtain the steady state condition equivalent to Eqn (5.9):

$$\frac{(k_0 + k_t)^2}{k_0}(a_0 - a_{ss}) = k_c\, a_{ss}(a_0 + b_0 - a_{ss})^2 \qquad (5.34)$$

or

$$\underbrace{\frac{(\kappa_0 + \kappa_t)^2}{\kappa_0}(1 - \alpha_{ss})}_{\mathcal{L}} = \underbrace{\alpha_{ss}(1 + \beta_0 - \alpha_{ss})^2}_{\mathcal{R}} \qquad (5.35)$$

In dimensionless terms, where $\kappa_t = k_t/k_c a_0$ and the other groups are as in Eqn (5.9). The only difference, then, is that the gradient of the flow line $\mathcal{L}$ depends on flow rate as $(\kappa_0 + \kappa_t)^2/\kappa_0$ instead of simply κ_0. This difference is not particularly significant at high flow rates; with $\kappa_0 \gg \kappa_t$, the slope is effectively given by κ_0 and so we find step flow lines at high flow rates as before. The different dependence of the slope on flow rate is, however, important at low flow rates, with the slope becoming infinite in the present case but tending to zero in Eqn (5.9) for the system without the termination step. The flow line has a minimum slope of $4\kappa_t$ in the present case, arising when $\kappa_0 = \kappa_t$. If this line of minimum slope lies above the tangent to the rate curve $\mathcal{R}$, which is independent of κ_t, then no tangencies will be possible, so no ranges of bistability will exist even if $\beta_0 < \frac{1}{8}$. If the line of minimum slope lies between the tangencies, the system will twice pass over the upper (extinction) line of tangency, giving rise to an isola pattern. Finally, if κ_t is sufficiently small, the flow line of minimum slope lies below the lower tangency

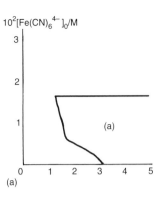

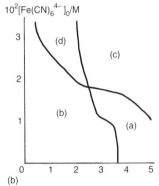

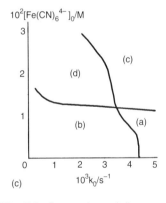

Fig. 5.8 Cross-shaped diagrams for the oscillatory Landolt (EOE) reaction in which $Fe(CN)_6^{4-}$ is added to the inflow: (a) $T = 20\ °C$, only bistability; (b) $T = 30\ °C$, bistability and oscillations; (c) $T = 40\ °C$, bistability and oscillations. (Reprinted with permission from Edblom *et al.* (1986), © American Chemical Society.)

66 *Bistability and oscillations in flow reactors*

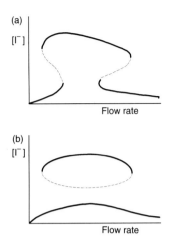

Fig. 5.9 (a) Mushroom and (b) isola responses for the steady-state concentration of iodide ion in the Landolt reaction with an additional inflow of solvent. (Reprinted with permission from N. Ganapathisubramanian and K. Showalter (1984). *J. Am. Chem. Soc.*, **106**, 816. © American Chemical Society.)

For full details see P. Gray and S.K. Scott (1990). *Chemical oscillations and instabilities: non-linear chemical kinetics*. Oxford University Press, Oxford, Chapter 6;. (1985). *J. Phys. Chem.*, **89**, 22–32: S.K. Scott (1987). *Acc. Chem. Res.*, **20**, 186–91.

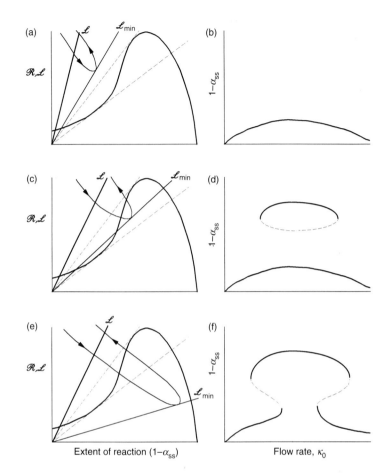

Fig. 5.10 The evolution of mushroom and isola patterns in the cubic autocatalysis model with termination: (a) and (b) the termination rate constant is high, so the line of minimum slope lies above the upper tangency, giving no bistability; (c) and (d) the line of minimum slope lies between the two tangencies giving rise to an isola pattern as k_0 is varied; (e) and (f) the line of minimum slope lies below the low tangency, leading to a mushroom pattern.

and $\mathcal{R}$ and $\mathcal{L}$ can become tangential four times, giving rise to a mushroom pattern. This sequence is indicated in Fig. 5.10(a–f). The decay step B → C also allows the cubic autocatalysis model to exhibit oscillations. Mushrooms, isolas, and oscillations are also exhibited by exothermic reactions when self-heating of the reacting mixture leads to thermal feedback.

6 Oscillatory reactions in flow systems

The gas-phase reaction between hydrogen and oxygen is a classic 'set piece' of undergraduate kinetics courses, illustrating the influence of branched-chain kinetics on the simplest of all combustion reactions. Chain branching provides feedback and this reaction also proves to be a classic nonlinear kinetic system. In closed vessels, the p–T_a ignition limits are evidence of *parametric sensitivity*: two systems with similar initial conditions but on either side of the limit evolve to final equilibrium states that are very similar, but in one case the reaction proceeds slowly whilst in the other it occurs explosively. In a flow system, oscillations and true steady state bifurcations emerge.

6.1 The $H_2 + O_2$ reaction in closed vessels

Ignition limits

The behaviour of $H_2 + O_2$ mixtures at subatmospheric pressures over the temperature range 700–800 K is summarized in Fig. 6.1. Two distinct modes of reaction are observed depending on the exact initial pressure and temperature: *slow reaction*, in which the rate of conversion of the reactants to the product H_2O may be imperceptibly low, and *ignition*, for which the reaction proceeds on a millisecond timescale with explosive force and a considerable temperature excursion. The locus of p-T_a conditions separating these two forms gives rise to the p-T_a *ignition limits*: the three branches of the curve are labelled as the first, second, and third limits in order of increasing pressure. The exact location of these limits depends on various factors: the vessel size, packing, and coating (these affect the first limit most) and the initial gas composition including inert gases (mainly affecting the second limit).

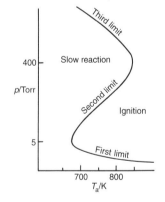

Fig. 6.1 Typical pressure–ambient temperature (p–T_a) ignition limits for the $H_2 + O_2$ reaction in a closed vessel.

Kinetic feedback mechanism

Although the reaction is strongly exothermic and self-heating can be detected in the slow reaction zone, the primary mechanism for the transition from slow reaction to ignition in the vicinity of the first and second limits ($p < 100$ Torr) is the branched-chain kinetics. Three overall processes can be distinguished: *initiation*, the *branching cycle*, and the *termination steps*.

Initiation processes, in general, are not easy to identify in branched-chain systems but for the present purposes the step

(0) $\quad\quad\quad H_2 + O_2 \rightarrow H + HO_2 \quad\quad$ Rate $= r_i = k_0[H_2][O_2]$

can serve as the route from the products to the first radical species. (In the subsequent analysis, HO_2 will be treated as a relatively unreactive species, even though it is formally a radical: the significant radical formed in Eqn (6.1) is the highly reactive H atom. The concentrations of the reactants H_2 and O_2 will also be considered as effectively constant.)

The branching cycle was discussed earlier (1.3). It comprises the following elementary steps:

(1) $\quad\quad\quad OH + H_2 \rightarrow H_2O + H$

(2) $\quad\quad\quad H + O_2 \rightarrow OH + O$

(3) $\quad\quad\quad O + H_2 \rightarrow OH + H$

These contribute to an overall branching process:

$$H + 3H_2 + O_2 \rightarrow 3H + 2H_2O \quad\quad \text{Rate} = r_b = k_2[H][O_2]$$

Chain branching competes with termination, which removes radicals from the system. In the vicinity of the lower limits, the important termination processes are:

(4) $\quad\quad\quad H \rightarrow \tfrac{1}{2}H_2$

(5) $\quad\quad\quad H + O_2 + M \rightarrow HO_2 + M$

The net termination rate is simply the sum of the rates of these individual steps:

$$r_t = k_4[H] + k_5[H][O_2][M] \tag{6.1}$$

Step (5) involves a *third body* species M. The role of M is to remove energy from the newly formed HO_2 complex and hence stabilize the product before it redissociates or reacts via the channel corresponding to step (2). Any species present in the reaction mixture can play this third body role, although the most important third bodies will be those present in the greatest concentrations (H_2, O_2, inert diluents and, in the later stages of the reaction, H_2O). We will consider the form of step (5) in more detail shortly.

Ignition limit condition

The rate at which the H atom concentration changes can be written in terms of the rates of initiation, branching, and termination

Formally, this equation can be obtained by evaluating the expressions for the steady-state concentrations of OH and O and substituting these into the rate equation for [H].

$$d[H]/dt = r_i + 2r_b - r_t = r_i + \phi[H] \tag{6.2}$$

where

$$\phi = 2k_2[O_2] - (k_4 + k_5[O_2][M]) \tag{6.3}$$

is the *net branching factor*.

For a given initial composition, pressure and temperature, r_i and ϕ are effectively constants and Eqn (6.2) can be integrated relatively easily to give

$$[H] = (r_i/\phi)(e^{\phi t} - 1) \tag{6.4}$$

(strictly, for $\phi \neq 0$). This exponential time dependence of the H atom concentration differs qualitatively for the two possible cases of $\phi < 0$ and $\phi > 0$.

$\phi < 0$: If the net branching factor is negative, i.e., if the termination rate exceeds the branching rate at all times, Eqn (6.4) can be rewritten as

$$[H] = [H]_{ss}(1 - e^{\phi t}) \tag{6.5}$$

where $[H]_{ss} = -r_i/\phi$ (which is a positive quantity for $\phi < 0$). This shows that [H] approaches its steady-state value, determined by the balance between production through the initiation process and removal by the net termination, as indicated by curve (a) in Fig. 6.2. The approach is exponential as the term $e^{\phi t}$ decays with $\phi < 0$. Because the initiation rate is typically very low, the steady-state radical concentration will also typically be low, unless ϕ is very close to zero. In general, then, the condition $\phi < 0$ corresponds to the case of slow reaction in the p–T_a diagram.

$\phi > 0$: Under conditions such that the branching rate exceeds the total termination rate, Eqn (6.4) indicates that the H atom concentration will grow exponentially and continuously in time, curve (b) in Fig. 6.2. Ultimately, the steady-state approximation and the assumption that the reactant concentrations can be treated as constants will break down, but this accelerating rate is clearly quite different from the exponential decay to a low steady radical concentration and, instead, corresponds to explosive behaviour.

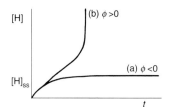

Fig. 6.2 Different evolution of the H atom concentration for systems with (a) $\phi < 0$ evolving to a steady state and (b) $\phi > 0$ accelerating to ignition.

Interpretation of ignition condition

The case separating slow reaction from ignition is now clearly that for which $\phi = 0$, i.e., for which the rates of branching and termination are exactly balanced. In full, this condition becomes

$$2k_2[O_2] = k_4 + k_5[O_2][M] \tag{6.6}$$

The rate constants are functions of the temperature, whilst $[O_2]$ and $[M]$ depend on the pressure. Equation (6.6) is thus an expression for the ignition limit pressure, p_{ign}, as a function of the ambient temperature, T_a. For the moment, we may approximate $[M]$, the total concentration of possible third bodies, by p_{ign}/RT_a and take k_4 to be independent of pressure. Equation (6.6) then has a quadratic form:

$$2k_2 x_{O_2}(p_{ign}/RT_a) = k_4 + k_5 x_{O_2}(p_{ign}/RT_a)^2 \tag{6.7}$$

which has either two or no positive real roots, depending on T_a (and, hence, the rate constants) and the mole fraction of oxygen, x_{O_2}. The two real roots would then give the ignition limit pressures for a given T_a (Fig. 6.3); as T_a is reduced, these two roots merge and become a complex pair at the junction of the first and second limits.

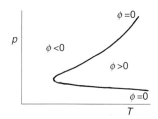

Fig. 6.3 The two roots of Eqn (6.14) define the first and second p–T_a ignition limits along which $\phi = 0$. To the left of the limit, $\phi < 0$ and there is slow reaction with branching controlled by termination; to the right of the limit, $\phi > 0$ and ignition occurs.

First limit

In fact, the termination step (4) is determined by the diffusion of H atoms to the walls and, hence, k_4 is likely to decrease as the pressure increases. This termination process will be most influential at the lowest pressures, appropriate to the first limit. At low pressures, the three body process (5) is also likely to be unimportant, so the condition for the first limit can be simplified to a balance between the branching cycle and termination step (4), giving

$$k_2 x_{O_2}(p_{ign}/RT_a) = k_4 \qquad (6.8)$$

At the lowest pressures, $p < p_{ign}$, the rate of the gas-phase branching process is reduced below that of the termination step; H atoms, on average, reach the wall and are removed before they react with an O_2 molecule. Termination controls the reaction and a steady state can be established. As the pressure is increased, so the branching rate increases proportionately and the termination rate decreases. The two rates become exactly equal at p_{ign} and for $p > p_{ign}$, the branching process wins out.

The rate constant k_2 is a sensitive function of the temperature as it has a relatively high activation energy, $E_2 \approx 70 \text{ kJ mol}^{-1}$; the termination process, on the other hand, will not change much with temperature. Thus p_{ign} will decrease as T_a increases. The nature of the vessel surface is important here, different surface coatings or surface-to-volume ratios will be reflected in different appropriate values for k_4. Vessels coated with 'aged' boric acid are almost incapable of effecting the termination step, so $k_4 \approx 0$ and the first limit effectively does not exist. With a more active surface, such as KCl, the first limit is raised to sufficiently pressures that it can be relatively easily located experimentally.

Second limit

As the pressure is increased, so the simple termination of H atoms at the wall via step (4) becomes less important. It is the competition between the second-order branching process and the third-order termination step (5) that determines the critical pressure for the second limit. The condition for a balance: above the limit, the termination step controls the reaction, but as the pressure is decreased the rate of this step decreases more quickly than that of the branching step. The two rates become equal when $p = p_{ign}$, given approximately from Eqns (6.6) or (6.7) by

$$2k_2 = k_5(p_{ign}/RT_a) \qquad (6.9)$$

Again, k_2 is a sensitive function of the temperature, increasing as T_a increases. The temperature dependence of k_5 is more subtle. The role of the third body is to remove energy and hence stabilize the produce relative to redissociation. The more thermal energy the species possesses, the greater the excess that needs to be removed. This makes stabilization harder to achieve at higher temperatures, so k_5 effectively *decreases* as T_a increases. The critical condition (6.9) thus indicates that p_{ign} should increase as the ambient temperature is raised.

To explain the influence of the initial gas composition and of inert gases on the location of the limit, we need to consider the termination process (5) more carefully. Various species present at reasonable concentrations can play the role of the third body M, but not all species are necessarily as efficient as each other in facilitating the required energy transfer. It turns out that the light molecule H_2 is approximately three times more efficient than the heavier O_2 or N_2 (which have approximately the same efficiency); the triatomic product H_2O has especially suitable energy levels into which the excess energy can be transferred effectively and so is approximately six times more efficient than H_2.

If we consider a mixture of H_2, O_2, and N_2 then the total termination process (5) actually occurs through the three steps

(5_{H2}) $H + O_2 + H_2 \rightarrow HO_2 + H_2$ Rate = $k_{t,H_2}[H][O_2][H_2]$

(5_{O2}) $H + O_2 + O_2 \rightarrow HO_2 + O_2$ Rate = $k_{t,O_2}[H][O_2]^2$

(5_{N2}) $H + O_2 + N_2 \rightarrow HO_2 + N_2$ Rate = $k_{t,N_2}[H][O_2][N_2]$

with $k_{t,H_2} \approx 3k_{t,O_2}$ and $k_{t,O_2} \approx k_{t,N_2}$. The termination rate then becomes

$$r_t = k_{t,H_2}[[H][O_2][H_2] + k_{t,O_2}[H][O_2]^2 + [k_{t,N_2}H][O_2][N_2] \quad (6.10)$$

This can be written in the form

$$r_t = k_{t,H_2}[H][O_2][M] \{x_{H_2} + a_{O_2}x_{O_2} + a_{N_2}x_{N_2}\} \quad (6.11)$$

where $[M] = p/RT_a$ is, as before, the total concentration of species in the gas phase. The reaction rate constant k_{t,H_2} is now specified as that with H_2 as the third body. The correction factor $\{\Sigma a_i x_i\}$ involves the mole fractions of the various species $i = H_2, O_2, N_2$, and the *third body efficiencies* $a_i = k_{t,i}/k_{t,H_2}$. By definition, the third body efficiency of H_2 is unity, whilst $a_{O_2} \approx a_{N_2} \approx \frac{1}{3}$ and, for later reference, $a_{H_2O} \approx 6$.

Because the a_i terms are not all equal to unity, r_t is now a function of the mixture composition. Replacing H_2 by a less efficient third body such as O_2 decreases the termination rate and, hence, alters the critical pressure (p_{ign} increases in this case so the second limit moves to lower temperatures, as indicated in Fig 6.4).

The nature of the surface can also play a role at this limit, even though the branching and termination steps have been portrayed as occurring in the gas phase. This arises through the subsequent fate of the HO_2 species. Although a radical, it is much less reactive than H, OH, and O and under the prevailing conditions it can reach the vessel walls unreacted. If the surface is efficient at removing HO_2, the chain will effectively end as soon as HO_2 is formed in step (5). With inefficient or reflective surfaces, such as aged boric acid, HO_2, is not removed and so its concentration can build up. Other reactions then become important, including the step

$$HO_2 + HO_2 \rightarrow H_2O_2 + O_2$$

If H_2O_2 remains intact, step (5) still constitutes a termination, but at higher pressures, H_2O_2 may be thermally dissociated via

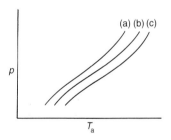

Fig. 6.4 The influence of composition, via third body efficiencies of H_2 and O_2, on the second ignition limit: (a) $H_2 + 2O_2$; (b) $H_2 + O_2$; (c) $2H_2 + O_2$. Increasing the concentration of the relatively more efficient third body H_2 shifts the limit to higher temperatures.

72 Oscillatory reactions in flow systems

$$H_2O_2 + M \rightarrow OH + OH + M$$

R.R. Baldwin and R.W. Walker (1972). In *Essays in chemistry* (ed. J.A. Barnard, R.D. Gillard, and R.F. Hudson), Vol. 33, pp. 1–37. Academic Press, London.

returning radicals to the pool. The mathematical analysis of the larger scheme appropriate to such vessel-coating is slightly different from that given above, but has the same essential feature of a low steady-state reaction rate giving rise to explosive growth as the pressure is deceased or the ambient temperature is increased.

6.2 H₂ + O₂ reaction in a CSTR

Ignition limit

The $H_2 + O_2$ reaction has also been studied in a well-stirred, flow reactor over a range of pressure and temperature similar to that for the second ignition limit above (10 < p < 100 Torr; 650 < T < 800 K). The system can again choose between slow reaction or ignition, and there is a sharp boundary in the form of an ignition limit between these. The slow reaction state is now a true steady state, with a balance between chemical kinetics and the inflow and outflow rates. On the ignition side of the limit, however, the system is substantially different from a closed reactor. In the closed system, ignition is a one-off event that leads to essentially complete conversion of the reactants to the product H_2O. In a flow reactor, however, the reactants will be replenished and the product removed. Two forms of behaviour are observed beyond the limit, as indicated in Fig 6.5. At higher pressures ($p > c.$ 40 Torr) a *steady flame* is supported; the ignited state is maintained at a steady rate exactly matching the inflow rate of the reactants. Because the system is operated with only moderate flow rates (the mean residence time is typically of the order of 2 to 10s) the maximum reaction rate is limited and

P. Gray, J.F. Griffiths, and S.K. Scott (1984). *Proc. R. Soc.* **A394**, 243.

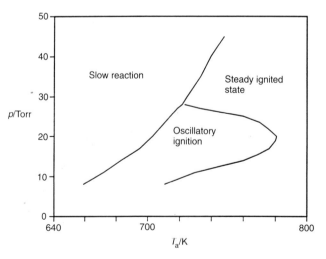

Fig. 6.5 The p–T_a limits for the $H_2 + O_2$ reaction in a well-stirred flow reactor. An ignition limit separates slow reaction from ignition, but two forms of ignition behaviour occur: oscillatory and steady flame.

so the 'flame' is accompanied by a relatively low increase in temperature, perhaps of the order of 10 to 20K.

Oscillatory ignition

At lower pressures, the system exhibits an oscillatory ignition. On crossing the limit, e.g., by increasing the ambient (oven) temperature, there is an ignition very similar to that observed in a closed vessel, with a transient gas temperature rise of several thousand kelvin. Complete consumption of the reactants occurs on a millisecond timescale. The reaction rate then falls to zero and the temperature falls back to the ambient. There is then a period of essentially no reaction, during which the reactant concentration builds up again and the product concentration decreases through the flow. Some time later, perhaps several minutes, the system has returned to a composition at which ignition develops again and the cycle is repeated. Example records of temperature rise, light emission, and species concentrations as functions of time in this oscillatory ignition region are shown in Fig. 6.6. These waveforms are appropriate to a system located just beyond the ignition limit, where the oscillations are sharp and separated by periods of quiescence. The thermocouple signal is strongly attenuated in the ignition event and so does not give a good estimate of the magnitude of the temperature excursion. A more accurate record is provided via the rotational fine structure on the OH signal determined by laser absorbance spectroscopy. This record, Fig. 6.7, also allows the OH concentration to be measured as a function of time during the oscillation. The H atom concentration can be followed using resonance-enhanced multiphoton ionization (REMPI), although absolute concentrations are not obtained in this way.

D.L. Baulch, J.F. Griffiths, W. Kordylewski, and R. Richter (1991). *Phil. Trans. R. Soc.*, **A337**, 199–210.

Self-inhibition

The feedback process leading to the development of ignition in the $H_2 + O_2$ system, based on the competition between branching and termination, has been discussed. A second aspect is important in the clockwork driving the oscillatory nature observed in flow reactors, namely *self-inhibition*. The product of the ignition event is the H_2O species. As mentioned previously,

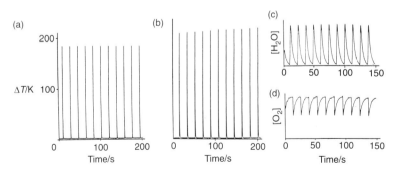

Fig. 6.6 Example experimental records for oscillatory ignition of (a) self-heating, $\Delta T = T_{gas} - T_a$; (b) light intensity I, (c) $[H_2O]$ and (d) $[O_2]$.

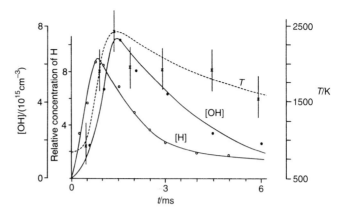

Fig. 6.7 Example records of oscillatory ignition from laser absorption and REMPI experiments. (Reprinted with permission from Baulch *et al.* (1991). *Phil. Trans. R. Soc.* **A337**, 199. © The Royal Society, London.)

this has a significantly higher third-body efficiency in the termination step (5) than the reactants H_2 and O_2 from which it is formed. Thus the change in composition that arises during the ignition gives rise to a change in the competition between branching and termination. At any given instant, the net branching factor can be calculated from

$$\phi = k_2[O_2] - (k_4 + k_{5,H_2}[O_2][M]\{\Sigma a_i x_i\} + k_f) \tag{6.12}$$

The flow rate k_f appears as another termination process as H atoms flow out of the reactor. In fact, under the conditions for which this reaction has been studied, neither k_4 nor k_f make a significant contribution to Eqn (6.12) and so the sign of ϕ is determined again by the competition between steps (2) and (5):

$$\phi/[O_2] = k_2 - k_{5,H_2}[M]\{\Sigma a_i x_i\} \tag{6.13}$$

Exactly on the limit, $\phi = 0$. As the slow reaction does not give rise to much consumption of the reactants, the mole fractions that are significantly in Eqn (6.13) at the limit are those corresponding to the inflow composition, e.g. $x_{H_2} = x_{O_2} = \frac{1}{2}$. For some slightly higher ambient temperature, k_2 will increase so ϕ becomes positive and the system develops into ignition.

After the ignition event, and once the temperature and pressure have recovered from their corresponding excursions, the composition of the system is quite different. Now the significant contributions to the term $\{\Sigma a_i x_i\}$ involve that from the product and any contribution from the reactant in stoichiometric excess. As a_{H_2O} is so much larger than a_{H_2} or a_{O_2}, the net branching factor for this new composition may be negative. This effectively prevents the inflowing H_2 and O_2 from reacting in the immediate post-ignition period. As the flow process slowly increases x_{H_2} and x_{O_2} and decreases x_{H_2O}, so ϕ increases again, passing through zero at some later stage at which the chemistry 'switches on' again. Effectively, the position of the ignition limit changes the composition: the system is at a point in Fig. 6.3

just to the right of the limit, as appropriate to the inflow composition. After the ignition, the system remains at the same point on the diagram, but, in the presence of the more efficient third body H_2O, the limit now jumps to the right, lying at higher T_a, so the system is now in the slow reaction zone appropriate to the new composition. As the flow process continues, so the limit drifts slowly back to the lower temperatures and an ignition follows as it passes over the p–T_a location of the system.

So, we see that the product suppresses further reaction, providing self-inhibition in this system. The period between ignition events is primarily determined by the flow, i.e., by the mean residence time of species in the reactor—but the oscillatory event is a combination of flow and kinetic feedback. Various other phenomena described earlier, such as bistability and birhythmicity, have been observed in this system.

S.K. Scott (1991). *Chemical chaos*, Chap. 9, pp. 284–95, Oxford University Press.

6.3 The CO + O₂ reactions

The reaction between CO and O_2 is as simple in stoichiometry as that between H_2 and O_2

$$CO + \tfrac{1}{2}O_2 \rightarrow CO_2$$

The reaction is slightly more exothermic, $\Delta H = -286$ kJ mol^{-1} compared with -247 kJ mol^{-1} for the $H_2 + \tfrac{1}{2}O_2$ reaction, but again the principle feedback arises through isothermal branched-chain kinetics competing with termination. The reaction is of considerable technical importance. With any hydrocarbon fuel, the initial reactions involve degradation of the initial molecular structure and the formation of CO. The subsequent oxidation of CO represents the major source of both heat evolution and the final production of CO_2. The product is intimately involved with the greenhouse effect contributing to possible global warming. Not surprisingly, therefore, the CO + O_2 reaction has as long a history of study as the $H_2 + O_2$ system. In contrast with the latter, however, the oxidation of CO has yielded to mechanistic understanding – and, indeed, even to a basic agreement over the observed phenomena – only slowly.

It appears doubtful that the CO + O_2 reaction can support any chemical feedback process in its own rights. The reaction is, however, extremely sensitive to minute traces of hydrogen or hydrogen-containing species such as CH_4 or H_2O. It is not clear that the reaction has been, or ever could be, studied in the absence of such impurities—and clearly in the context of hydrocarbon oxidation, substantial sources of H will be present. The branching and termination kinetics of the $H_2 + O_2$ system can, thus, operate here too.

Closed vessel studies

The reaction between CO and O_2 can support chemiluminescence as the product CO_2 can be formed in an electronically excited state. *Steady glow*

M. Prettre and P. Laffitte (1929). *C. R. Acad. Sci.*, **189**, 177; P.G. Ashmore and R.G.W. Norrish (1951). *Nature*, **167**, 390.

76 *Oscillatory reactions in flow systems*

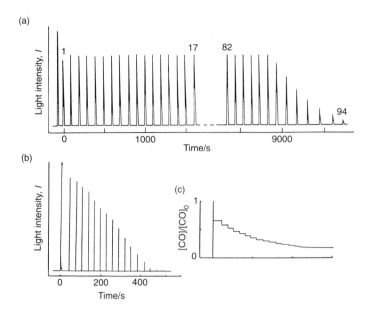

Fig. 6.8 Examples of long and short trains of oscillatory glow for CO + O$_2$ in closed vessels showing emitted light intensity and CO consumption.

was observed by Prettre and Laffitte in 1929, whilst Ashmore and Norrish observed long-lived sequences of *oscillatory glow*, with up to 100 pulses of light emission, in 1939, in a closed vessel (these results were not published until 1951). Ashmore and Norrish's reactor had previously been used to study the inhibiting effects of chloropicrin (CCl$_3$NO$_2$). Cleaning the reactor surface removed the oscillatory response, but this could be recovered by exposing the surface to the inhibitor again. The reaction is sensitive to the surface condition as well as to hydrogen. Other studies of the oscillatory glow (also termed the *lighthouse effect*) showed that suitably aged surfaces, without inhibitors, can support this mode. An example record is shown in Fig. 6.8. The emission, either steady or oscillatory, ceases well before complete reactant consumption.

Steady and oscillatory glow are alternatives and appear to occur at exactly the same p–T_a and composition conditions, being distinguished only by the state of the reactor surface. In a sequence of repeated identical experiments, the reaction will initially support steady glow; as the surface becomes suitably developed, so steady evolution gives rise to runs that exhibit oscillatory character; after, say, 100 experiments the oscillatory nature is lost again and all subsequent repetitions show just steady glow unless the reactor retreated with acid-washing. Oscillatory glow is most easily obtained in 'dry systems', from which as much hydrogenous matter as possible has been removed, e.g., by storage of the reactants over liquid O$_2$. The oscillations can, however, be observed even in the presence of up to 0.15% H$_2$, but with increasing 'wetness' and ignition peninsula due to the

J.R. Bond, P. Gray, J.F. Griffiths, and S.K. Scott (1982). *Proc. R. Soc.*, **A381**, 293.

Oscillations, Waves and Chaos

$H_2 + O_2$ system also appears, moving to lower temperatures as the H_2 concentration increases and so invading the glow peninsula.

Flow reactor studies

In continuous flow reactors, the steady and oscillatory glow responses become indefinitely sustained. In this configuration, these modes appear to arise under different p–T_a conditions, as indicated in Fig 6.9(a,b). If H_2 is added to the inflowing reactants, an oscillatory ignition mode can also be sustained. Again, this resides within a peninsula bounded by first and second limits, that moves across the p–T_a plane as the H_2 concentration increases, invading the regions of glow as indicated in Figs 6.9(c–e). Oscillatory ignition is distinguished from oscillatory glow by the extent of the fuel (CO) consumed in the light emission stage (virtually complete in ignition, possibly undetectably small for glow), the intensity of the emission and the temperature excursion (from several hundred to thousand kelvin for ignition, from less than ¼ to 20K for glow). In each case, there appears to be complete consumption of H_2.

P. Gray, J.F. Griffiths, and S.K. Scott (1985). *Proc. R. Soc.*, **A397**, 21.

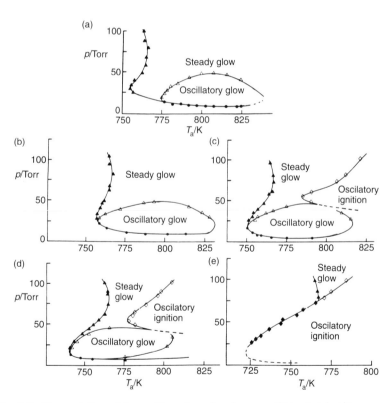

Fig. 6.9 The variation of the p–T_a regions for slow reaction, steady glow, oscillatory glow, and oscillatory ignition for the CO + O_2 reaction in a flow reactor in the presence of H_2: (a) no added H_2; (b) with 150 ppm H_2; (c) with 1500 ppm H_2; (d) with 7500 ppm H_2; (e) with 10% H_2.

Mechanistic interpretation

Accepting that the branching and termination steps (0–5) of the $H_2 + O_2$ system are of importance, the number of additional reactions arising from the CO is somewhat limited. The step

(6) $\qquad CO + OH \rightarrow CO_2 + H$

is significant as the major route from CO to CO_2 in this, and other systems but as a simple propagation step does not alter the radical pool. A second source of CO_2 is via the step

(7) $\qquad CO + HO_2 \rightarrow CO_2 + OH$

which has significance as it converts the unreactive HO_2 into the reactive radical OH and hence would weaken the net termination via step (5).

Perhaps the most significant addition is the step

(8) $\qquad CO + O + M \rightarrow CO_2^* + M$

The product here is a triplet state molecule (due to the triplet character of the ground state O atom) and hence this provides the source of the chemiluminescence. More importantly, this is a termination step involving O atoms. At low H_2 concentrations, this can compete with step (3). The latter is an important link in the branching cycle.

The details of the kinetic basis for oscillatory glow have not been fully established, but a qualitative picture appears to be emerging. For a true ignition, step (8) remains unimportant throughout the whole event. The net branching factor based on steps (2) and (5) becomes positive as, say, the ambient temperature is increased and an ignition develops. This ceases as the product builds up and the reactants (H_2 and CO) are extensively consume. The self-inhibition from H_2O and also of CO_2 prevents the inflowing gases from reacting immediately, as described for the $H_2 + O_2$ system. For oscillatory glow, reaction (8) participates. As the initial reaction develops, H_2 and CO are both consumed, but if CO is in great excess over H_2 the fall in concentration of the former is proportionately less. As the CO/H_2 ratio increases dramatically, so the competition for O atoms swings in favour of the termination step (8) away from the branching step (3). The radical concentrations will fall again, without significant consumption of the major fuel and, hence, without significant release of heat. The CO/H_2 ratio will fall again with the subsequent net inflow of fresh H_2 and the process can repeat.

Such 'hand-waving' explanations of oscillations often sound seductively reasonable, only to fail quantitative test. However, computations on small kinetic schemes involving the above steps with recommended values for the rate constants confirm the basic form of this account. The explanation must be adapted to encompass oscillatory glow in closed systems, where there is no replenishment of the H_2 concentration by flow. A possible route is the water gas shift process

$$CO + H_2O \rightleftharpoons CO_2 + H_2$$

This might be established via heterogeneous reactions on the vessel walls, helping to account for the sensitivity of the glow to the state of the surface in closed reactors—an effect that appears much less apparent in flow systems (the required experimental conditions for oscillatory glow are also somewhat different in the two types of system).

Development of oscillations: chaos

For equimolar mixtures of CO and O_2 containing ½%H_2, the ignition limit has moved to sufficiently low ambient temperatures so as to cover the regions of oscillatory and steady glow for pressures less than 50 Torr, as indicated in Fig. 6.10. Just at the right of the ignition limit, the system displays oscillatory ignition that have the 'relaxational' form described earlier, with sharp spikes separated by periods of replenishment accompanied by virtually no reaction. Each ignition has exactly the same amplitude and period as its predecessor, i.e., there is a *period-1* waveform. At the highest ambient temperatures, the oscillations give way to a steady flame. If the reactor is operated at the highest or lowest pressures indicated in the figure, the oscillations retain their period-1 character across the whole region of oscillatory ignition, simply decreasing in amplitude as T_a increases. For intermediate pressures, 15 < p/Torr < 45 roughly, there is a more interesting sequence. The development of the waveform as T_a is allowed to increase slowly, so as to traverse the region of *complex oscillations* indicated in Fig. 6.10 is illustrated for two cases in Fig. 6.11. Trace (a) corresponds to a relatively high pressure within this range, p = 40 Torr, which just catches the upper tip of this region. Over a narrow range of temperature, centred on T_a = 803.3K, a different, *period-2* waveform is observed in which there is one large and one small excursion in each repeating unit. As the period

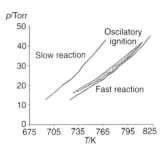

Fig. 6.10 The p–T_a regions for slow reaction, oscillatory ignition, and steady flame for CO + O_2 with 1% H_2 in a flow reactor. The hatched region corresponds to conditions for complex oscillations.

B.R. Johnson and S.K. Scott (1990). *J. Chem. Soc. Faraday Trans.*, **86**, 3701; (1991). *Chaos*, **1**, 387.

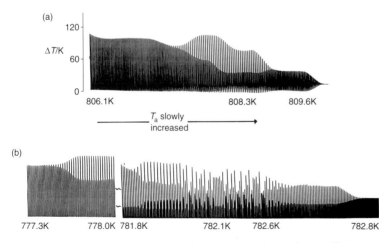

Fig. 6.11 Representative scans through the region of complex oscillations made by allowing the ambient temperature to increase slowly at fixed pressure: (a) p = 40 Torr showing a region of period-2 oscillations; (b) p = 20 Torr showing various waveforms.

80 *Oscillatory reactions in flow systems*

between each excursion, of either size, is roughly the same as that between each excursion in the period-1 mode, the total period becomes doubled: this change from period-1 to period-2 is known as a *period-doubling*. The figure also shows how the difference in amplitude grows smoothly as the system enters this region: there is no sharp jump immediately from period-1 to period-2 with a clear large–small character. This is, thus, a *supercritical period-doubling bifurcation*. As T_a rises further, the difference in amplitude decreases again and there is the reverse process—*a period halving*—to yield small amplitude period-1 oscillations. As the ambient temperature increases further, the period-1 oscillation gives way to a steady state via a supercritical Hopf bifurcation (see Section 5.6).

The scan in Fig. 6.11(b) corresponds to a lower pressure, $p = 20$ Torr, so the system passes through the widest section of the complex oscillation region. There is, effectively, more room for additional changes in the waveform and, specifically, for more period doublings. A period-4 solution can be seen clearly emerging from the period-2. There is then a region of very

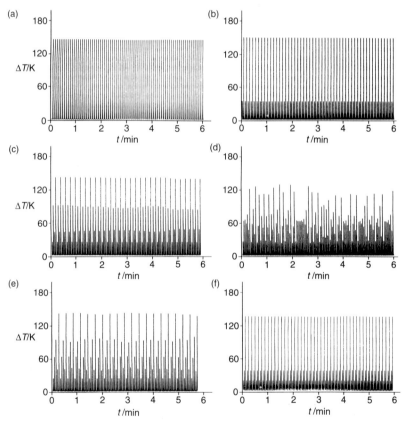

Fig. 6.12 Time series (temperature excess, ΔT) captured for steady experimental conditions with $p = 20$ Torr: (a) period-1; (b) period-2; (c) period-4; (d) chaotic response; (e) period-5; (f) period-3. This sequence is observed as the ambient temperature is increased stepwise.

complex structure, before period-4, then period-2 and finally period-1 emerge as T_a sweeps through this region. In order to unravel the complex structure between the period-4 states, the experiments must be repeated with the ambient temperature allowed to stabilize at various values within this range. The 'post-transient' oscillations obtained under these conditions, some of which are presented in Fig. 6.12, show that additional period doublings to yield period-8, period-16, and higher periodicities follow very closely on each others heels after the period-4. The range of T_a over which these can be observed decreases as the periodicity increases. This *accumulation of period-doubling bifurcations* allows the system to develop a waveform with an infinitely-long repeating unit, such as is shown in Fig 6.12(d). Such *aperiodic* or *chaotic* responses have some unfamiliar properties, but are simply a natural consequence of chemical feedback.

6.4 Characteristics of chaos

The first task when faced with a complex time-dependent response is to distinguish between genuine chaos and experimental noise. Although such noise will generally have small amplitude, the chemical nonlinearities can produce an amplification effect. A simple test between these possibilities is to construct a *next-amplitude map* from the experimental data. If we plot the peak temperature rise for the first excursion in Fig. 6.12(d) against the peak temperature of the second, then the peak temperature of the second against that of the third and so on, these $\Delta T_{\max,n}$ vs. $\Delta T_{\max,n+1}$ pairs lie as shown in Fig. 6.13. The points do not lie randomly distributed over the whole diagram, as they would if stochastic noise were driving the complexity. Instead they seem to lie on a relatively simple, single-humped curve exhibiting a maximum. This, in turn, suggests that these is some rule determining the size of the next ignition in terms of the amplitude of the most recent ignition, hence the phrase *deterministic chaos*. (Similar next-amplitude maps can be formed in the same way for the period-*n* responses: these simply give rise to a discrete set of *n*-points between which the system jumps as it evolves.)

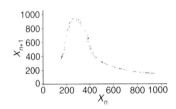

Fig. 6.13 Next-maximum map, plotting digitized thermocouple record at the peak temperature excess for one ignition against the same record for the next ignition successively through the time series.

Another approach frequently used to test for ordered structure within apparent chaos is the method of *reconstructing attractors*. The experimental thermocouple record consists of a series of measurements related to the temperature rise in the reacting gases taken at regular time intervals (e.g., every 10 ms): $\Delta T(t_1)$, $\Delta T(t_2)$, $\Delta T(t_3)$, etc., where $t_n = n(\Delta t)$ and Δt is the sampling interval. This record can be used to obtain a sequence of *x*–*y* pairs using the method of *time delay*. We choose a delay time $t_d = m(\Delta t)$ where m is some integer; the *x*–*y* pairs are now formed by taking $\Delta T(t_i)$ and $\Delta T(t_i + t_d)$. So, for example, if we choose a sampling time of 10 ms and a delay time of 100 ms, so $m = 10$, we take the following pairs: ($\Delta T(t_1)$, $\Delta T(t_{11})$), ($\Delta T(t_2)$, $\Delta T(t_{12})$), ($\Delta T(t_3)$, $\Delta T(t_{13})$) and so on. Performing this for the data corresponding to the various responses in Fig. 6.12 yields the attractors shown in Fig. 6.14. For the period-*n* responses closed circuits with *n*-loops emerge. These period-*n* limit cycles (see Section 3.4) correspond to

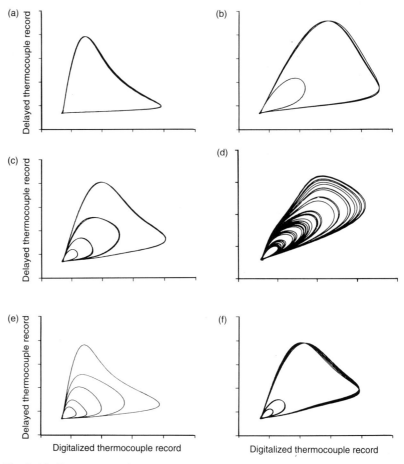

Fig. 6.14 Reconstructed attractors, shown in two-dimensional projection, for the time series in Fig. 6.12 using the delay method.

the *asymptotic trajectories* to which the system is 'attracted; and on which it continues to move for as long as the experimental conditions are maintained constant.

The attractor emerging from the aperiodic time series has a more complex form, Fig. 6.14(d), but essentially still comprises a series of circuits (there will be an infinite number). In fact the circuits finally add up to provide something more than a simple one-dimensional structure. The period-n limit cycles are all one-dimensional objects (lines), where as this *strange attractor* actually has a dimension that is greater than 1: it is a *fractal* and has a non-integer dimension slightly greater than two.

The major aspect of chaotic evolution or 'motion on a strange attractor' that is so different from more familiar responses arises when the evolution of two very similar, but slightly different, systems are compared. These systems may have exactly identical experimental conditions, in terms of p, T_a, mixture composition, etc., but vary to some small extent in their initial conditions. Such small differences are inevitable in any experimental or 'real' context as exactly identical systems to the finest level of detail can never be reproduced. Under experimental conditions such that the systems evolve from their initial states to (the same) final steady state, the small differences between them *decrease* (usually exponentially) in time. This is why the scientific approach of repeating experiments and taking means and standard deviations has become so well established.

A similar convergence from different, but similar, initial conditions arises when the system evolves to a period-n limit cycle. Under these conditions, however, some residual difference may remain, as a difference in phase in the oscillations of the two systems. For chaotic systems, however, small initial differences *increase* exponentially in time; two systems, no matter how similar in their initial conditions, grow less similar in time. The rate at which this growth in disparity occurs depends on the given system, but is a characteristic feature of all chaotic responses. In this situation, the long-term prediction of the behaviour of a given system is also impossible—not because of ignorance, but as a real matter of fact. Only if the initial conditions (i.e. the initial concentrations and temperature) of the system could be specified to infinite precision could the real evolution be predicted. This also assumes that we have an exact chemical mechanism and infinitely precise values for the rate constants, their temperature dependence, and the experimental conditions. As the 'problem' lies within the fundamental properties of nonlinear feedback, spending extra money on a more precise determination of rate constants, for instance, does not help. Worse still, the improvement in predictability increases only logarithmically with increasing precision to which the initial conditions are known.

All is not lost, however; what is needed is a new way of approaching the problem. Means and standard deviations, and multiple repetitions of experiments are not the appropriate route. Various quantitative methods exist for assessing how long a chaotic system will remain predictable, *within certain acceptable limits*. Thus, we might be able to predict the future evolution of a given system with ±10% accuracy for 2 hours, say, or with ±20% accuracy for 2.9 hours. The *predictability horizon* depends on how well we can know the initial conditions, how fast similar systems diverge, and on how much we are prepared to tolerate uncertainty. Ultimately, however, the responses of two non-exactly identical systems become completely uncorrelated. This is the message from chaos.

6.5 Hydrocarbon oxidations: cool flames

Closed vessel behaviour

Oscillations and ignitions arise in the combustion of hydrocarbon fuels and have considerable technological significance. Alkane species and their derivatives (alkenes and partially oxygenated products such as aldehydes) react spontaneously with O_2 at temperatures around 400–500K—although methane requires a significantly higher temperature. In closed systems, there is an oscillatory mode known as *cool flame* behaviour. In a cool flame,

J.F. Griffiths (1986). *Adv. Chem. Phys.*, **64**, 203; J.F. Griffiths and S.K. Scott (1987). *Prog. Energy Combust. Sci.*, **13**, 161.

a proportion of the reactants are consumed and converted to partially oxygenated products including peracids, alcohols, and peroxides. There is a faint chemiluminescent emission in the blue region of the spectrum from electronically excited HCHO. This reaction event is accompanied by a temperature excursion of between 10 and 200K, typically. Sequences of up to ten such cool flames may be observed, but the effects of reactant consumption are pronounced and prevent this mode of reaction from being sustained to any great extent. Cool flames are quite different from true (one-off) ignitions, in which the major products are CO, CO_2, and H_2O and the temperature rise is of the order of 2000K.

Behaviour in a CSTR

In flow reactors, the cool flame mode can be sustained indefinitely, allowing it to be studied free from imposed transient effects. Some typical concentration histories for ethanal oxidation are shown in Fig. 6.15. Cool flames occur in a region of the p–T_a ignition diagram distinct from that for true ignition, as indicated in Fig. 6.16. There are also regions corresponding to two different steady state forms of reaction and a region of *complex ignition*.

At the lowest vessel temperatures, the reaction sits in a steady state of low extent of conversion of the reactants. A steady-state temperature rise can be measured, and this increases as the vessel temperature increases. At the highest vessel temperatures is a second region of steady-state reaction, but this now corresponds to high extents of reactant consumption and so is akin to the steady-flame state in the $H_2 + O_2$ reaction. In this region, the steady-state-temperature excess *decreases* with increasing vessel temperature. This remarkable phenomenon is termed the *negative temperature coefficient* or *n.t.c.*

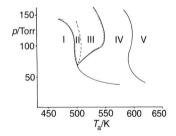

Fig. 6.16 The p–T_a diagram for ethanal oxidation in a CSTR showing the regions of I slow reaction, II two-stage ignition, III multi-stage ignition, IV cool flames, and V steady glow. (Reprinted with permission from P. Gray, J.F. Griffiths, S.M. Hasko, and P.G. Lignola (1981). *Proc. R. Soc.* **A374**, 313. The Royal Society, London.)

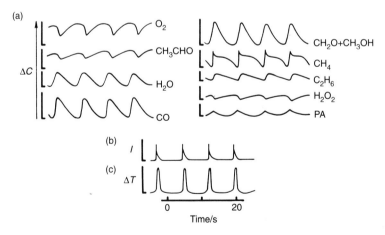

Fig. 6.15 a) Representative concentration, (b) temperature excess, and (c) emitted light intensity records for cool flame oxidation of ethanal in a CSTR. (Reprinted with permission from P. Gray, J.F. Griffiths, S.M. Hasko, and P.G. Lignola (1981). *Proc. R. Soc.* **A374**, 313. © The Royal Society, London.)

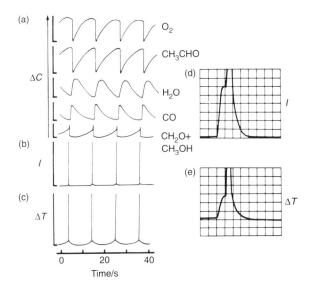

Fig. 6.17 (a) Representative concentration, (b) temperature excess, and (c) emitted light intensity records for ignition in the oxidation of ethanal in a CSTR: the enlargements of the ΔT and I traces (d,e) show the two-stage character of the ignition event. (Reprinted with permission from P. Gray, J.F. Griffiths, S.M. Hasko and P.G. Lignola (1981). *Proc. R. Soc.*, **A374**. 313. © The Royal Society, London.)

The waveform corresponding to the ignition response shows some important fine structure, as indicated in Fig. 6.17. As a precursor to the actual 'explosion event', there is a slower first stage which has many similarities to the evolution of the system in the cool flame region. The transition from the first to the second stage in the ignition process produces a shoulder in the emission intensity and temperature rise records as indicated. The complex or *multi-stage ignition* region supports two-stage ignition events preceded by distinct cool flame events. Figure 6.18 shows the temperature rise record for a four-stage ignition which has a repeating unit comprising two cool flames followed by a two-stage ignition. The number of cool flames between each two-stage ignition increases as the vessel temperature is increased through this region until the boundary with the cool flame region is reached.

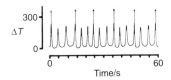

Fig. 6.18 Temperature excess record for a four-stage ignition. (Reprinted with permission from P. Gray, J.F. Giffiths, S.M. Hasko, and P.G. Lignola (1981). *Proc. R. Soc.*, **A374**, 313. © The Royal Society, London.

Thermokinetic mechanism

The mechanism for this behaviour involves crucially both chemical and thermal feedback. At the heart of the oscillatory clockwork is an equilibrium step involving a radical species, $R^\cdot$ and the O_2 molecule

(9) $$R^\cdot + O_2 \rightleftharpoons RO_2^\cdot$$

This equilibrium is highly temperature-sensitive because of the exothermic nature of the bond formation process. At low temperatures, the equilibrium lies in favour of the product $RO_2^\cdot$. Subsequent reaction of the RO_2 species lead to a series of chain-branching steps and, hence, to an acceleration of the overall rate of oxidation. These subsequent oxidation processes are significantly exothermic and so there is an increasing rise in the temperature of the reacting gas mixture. As the temperature increases, so the equilibrium in step (9) swings back in favour of the reactants. The system then changes to reactions involving R˙ rather than $RO_2^.$. These reaction tend to involve chain termination steps, so decreasing the overall rate of oxidation and, hence, the rate of heat evolution. If the reacting mixture cools sufficiently, the equilibrium may switch back in favour of the $RO_2^.$ species and allow branched-chain acceleration again, followed by a switch back to R˙ chemistry as the temperature increase. This combination of positive chemical feedback and effective thermal inhibition accounts both for the existence of cool flame oscillations and of the n.t.c.

Chemical considerations

The simplest realization of the thermokinetic mechanism described above arises with fuel molecules such as ethanal, in which a moderately reactive site exists, the —CHO hydrogen atom in this case. The reactions following on from step (9), in which R is CH_3, are then

(10) $\qquad CH_3O_2 + CH_3CHO \rightarrow CH_3O_2H + CH_3CO$

The radical CH_3CO can propagate the chain further (decomposing to CH_3 and CO), whilst CH_3O_2H also dissociates

(11) $\qquad CH_3O_2H \rightarrow CH_3O + OH$

to produce two more radicals. The increase in radical concentration through this subsequent step is termed *degenerate branching* to distinguish it from the direct branching in steps (2) and (3).

The channel involving methyl radical when the equilibrium in step (9) lies to the left is simply the recombination process leading to the formation of ethane, $CH_3 + CH_3 \rightarrow C_2H_6$.

For more general hydrocarbon systems, the H atom abstraction in step (10) has a significant activation energy ($E > 60$ kJ mol^{-1} compared to $E = 44$ kJ mol^{-1} for ethanal) and hence would proceed too slowly to support the cool flame phenomena. Instead, the general mechanism from RO_2 involves *intramolecular H atom abstraction*. Step (9) can be represented as

(12)
$$-CH-CH_2-CH_2- + O_2 \rightarrow \overset{\overset{\displaystyle O-O}{|}}{-CH}-CH_2-CH_2-$$

This then allows abstraction by the O atom of the H-atom on a β-C via the formation of a sterically favoured, six-membered ring

(13)
$$\overset{\overset{\displaystyle O-O}{|}}{-CH}-CH_2-\overset{\overset{\displaystyle H}{|}}{CH}- \rightarrow \overset{\overset{\displaystyle O_2H}{|}}{-CH}-CH_2-CH-$$

A second O_2 molecule can add to the new radical site, and this can then abstract the activated H atom from the carbon atom that had the original radical site, again via a six-membered ring. The resulting radical species will then dissociate:

(14)
$$-\underset{\underset{\text{O}_2\text{H}}{|}}{\text{C}}-\text{CH}_2-\underset{\underset{\text{HO}_2}{|}}{\text{CH}}- \rightarrow -\text{CO} + \text{CH}_2\underset{\text{O}}{\overset{}{\diagdown\diagup}}\text{CH}- + 2\text{OH}$$

The opening of the oxide ring may well then follow. In this way, the internal molecular structure of the fuel can play a significant role in determining whether suitable β-C sites exist, etc.

The spontaneous reactions of hydrocarbon species of the above type are significant in the phenomenon of *engine knock*. In a spark ignition engine, the flame front initiated by the spark at the optimum moment towards the end of the compression stroke propagates through the combustion cylinder. This front compresses the unreacted gas ahead, which has already been heated to some extent in the compression stroke itself. This additional heating can initiate the spontaneous reactions on a timescale sufficiently short to occur before the flame fully traverses the cylinder. The resulting reduction in the energy content arising from the partial oxidation and pressure fronts induced cause inefficiency in the combustion process and mechanical events that give rise to the acoustic 'knocking'. Knowledge of the underlying chemical cause of this spontaneous reaction is necessary to improve the resistance of fuels to knock as conventional antiknocks such as tetraethyl lead are withdrawn.

6.6 BZ reaction in a CSTR

The BZ reaction has been widely studied in well-stirred flow reactors. For moderate flow rates, the reaction exhibits large amplitude oscillations very similar to those observed in closed systems (Chapter 3), but with the oscillations now being indefinitely sustained and each excursion having exactly the same amplitude and period as the previous one; the slow variations in the background concentrations of the reactants and the build-up of product concentrations is now prevented by the continuous flow process. More complex oscillations emerge, however, at both low and high flow rates.

Low flow rate behaviour

As the flow rate is reduced in the BZ system, so the oscillatory response changes from the simple, large amplitude period-1 waveform just described. The system undergoes a period-doubling cascade similar to that presented for the CO + O_2 system in Section 6.3. This gives rise to a range of flow rates over which the system exhibits chaos. Next maximum maps in this region show the same simple hump shape as indicated in Fig. 6.13 and the consequences of this chaos are the same as that for the CO + O_2 system. If the flow rate is decreased further, there is a period-halving sequence that produces a small amplitude period-1 solution. At yet lower flow rates, the oscillations give way to stable steady-state behaviour and this steady-state composition

J.-C. Roux, R.H. Simoyi, and H.L. Swinney (1983). *Physica*, **8D**, 257.

approaches the (closed vessel) equilibrium composition smoothly as the flow rate is reduced to zero.

High flow rate behaviour

At high flow rates, the large amplitude, period-1 solution also becomes more complex. The sequence of increasing complexity is indicated in Fig. 6.19. This change in waveform is sometimes, and rather inaccurately, termed *period-adding*. In fact, the sequence sees an increase in the number of individual excursions that go to form a total repeating unit, so might be better termed *maximum-adding or peak-adding*. The basic sequence is to go from one large peak (Fig. 6.19a) to a repeating unit consisting of one large and one small peak (a 1^1 state, Fig. 6.19c) and subsequently to one large and two small (1^2, Fig. 6.19f) and so on through 1^3, 1^4, etc. Eventually, the number of small peaks becomes effectively infinite, so there are no large peaks in the repeating unit, giving a 1^∞ or 0^1 state, i.e., a small amplitude period-1 oscillation. In between these *parent* states are *concatenations*, consisting of mixed parent states. Thus between the 1^0 and 1^1 states is a range of flow rates for which a $1^0 1^1$ state is observed. This state, which can also be denoted 2^1 has two large and one small peak per repeating unit, Fig. 6.19(b). Further levels of concatenation can be found on closer inspection. A sequence between the 1^2 and 1^3 states consisting of $(1^2)^2 1^3$, i.e., a repeating unit consisting of one large and two small repeated twice followed by one large and three small, $1^2 1^3$, $1^2(1^3)^2$, $1^2(1^3)^3$, $1^2(1^3)^4$ and $1^2(1^3)^5$ as the flow rate increases has been reported.

In between some of these higher levels of concatenation, the system may show mixing of parent units that does not occur in a periodic way. For instance, in Fig. 6.19(e) the 1^1 and 1^2 appear aperiodically. Similar chaotic mixing is observed in the concatenations between the 1^2 and 1^3, and the 1^3 and 1^4 parents, Fig. 6.19(g) and (i) respectively.

Mechanistic interpretations of chaos in BZ reaction

Much of the modelling of the BZ reaction in Chapters 3 and 4 made use of a two-variable version of the Oregonator scheme. Models with two independent species can reproduce simple oscillations, but cannot give rise to complex oscillations or chaos: a minimum of three independent species is needed. The first numerical studies of the BZ reaction in a flow system used enlarged versions of the FKN mechanism, either including more species, such as HOBr, as participants in the chemistry rather than just products, or allowed the stoichiometric factor f to vary with the instantaneous composition in the reactor. The studies quickly produced the complex concatenated periodic states described above. However, these models based on the original Oregonator form typically failed to produce the aperiodic behaviour reported in experiments. This led to the opinion that this 'high flow rate' chaos might be an artifact of the experiments. Various causes were suggested and explored. A real experiment might have an imperfectly controlled flow rate, especially if peristaltic pumps are used. The apparently chaotic responses might arise because the system is continuously being moved back

J. Maselko and H.L. Swinney (1986). *J. Chem. Phys.*, **85**, 6430.

F.W. Schneider and A.F. Munster (1991). *J. Phys. Chem.*, **95**, 2130.

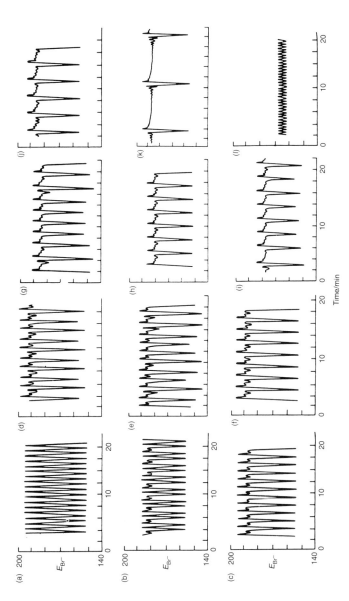

Fig. 6.19 Complex oscillations and chaos at high flow rates for the BZ reaction in a CSTR showing concatenations between states of large and small amplitude peaks. (Reprinted with permission from J.L. Hudson, M. Hart, and D. Marinko (1979). *J. Chem. Phys.*, **71**, 1601. © American Institute of Physics.)

and forth over a range in which various periodic states occur close together, so that different states are picked out almost randomly at different times. Another possibility is that the mixing within the reactor will be less than perfect. In the vicinity of the inlet ports, especially, local regions that vary significantly in compositions from the bulk of the reactor may arise. Oscillations in these regions might then have a different natural period and waveform and might 'force' the behaviour in the bulk of the reactor by imposing an effectively time-dependent inflow composition.

The simple next-maximum map approach can be used to distinguish between noise and chaos, as indicated in Section 6.5, and the aperiodic traces in Fig. 6.19 do indeed show simple humped maps, confirming chaos not noise. Such a test cannot prove that the chaos arises from the chemistry, however. If the chaos arose from fluid turbulence, this *universal property* of chaos would still emerge. Only recently has a chemically-sound model been derived that reproduces the observed sequences including the interposed chaotic states on the concatenations. Györgyi and Field have produced a set of models that, using either steady state or quasi-equilibrium assumptions, can be reduced to a four-species scheme that involves the original Oregonator variables [$HBrO_2$], [Br^-], and [M_{ox}] augmented by [BrMA], where BrMA is bromomalonic acid and is formed rapidly from HOBr. The model can then be written as:

L. Györgyi and R.J. Field (1992). *Nature*, **355**, 808.

(GF1) $\quad BrO_3^- + Br^- \rightarrow HBrO_2 + BrMA$

(GF2) $\quad HBrO_2 + Br^- \rightarrow 2BrMA$

(GF3) $\quad BrO_3^- + \tfrac{1}{2}HBrO_2 \rightleftharpoons HBrO_2 + M_{ox}$

(GF4) $\quad 2HBrO_2 \rightarrow BrO_3^- + BrMA$

(GF5) $\quad M_{ox} \rightarrow$

(GF6) $\quad BrMA + M_{ox} \rightarrow Br^-$

(GF7) $\quad BrMA \rightarrow Br^-$

The autocatalysis of Process B appears as the reversible step (GF3), limited by step (GF4) as before, and Process A is still represented by steps (GF1 and 2). Process C, the clock resetting, is now modelled by steps (GF5–7). The competition between these three steps governs how many bromide ions are produced for each M_{ox} reduced in a more explicit way than in the original Oregonator, where the curious stoichiometric factor f is used. Rather than 'guestimating' a suitable value for f, this approach seeks the values of the rate constants for the final three steps.

Further reading

Review articles

De Kepper, P., Boissonade, J., and Epstein, I.R. (1990). Chlorite-iodide reaction: a versatile system for the study of nonlinear dynamical behaviour. *Journal of Physical Chemistry*, **94**, 6525–36

Epstein, I.R., Kustin, K., De Kepper, P., and Orban, M. (1983). Oscillatory chemical reactions. *Scientific American*, **248**, 96–108.

Epstein, I.R. (1984). Complex dynamical behavior in 'simple' chemical systems. *Journal of Physical Chemistry*, **88**, 187–92

Gray, P. and Scott, S.K. (1985). Sustained oscillations and other exotic patterns of behavior in isothermal reactions. *Journal of Physical Chemistry*, **89**, 22–32.

Noyes, R.M. (1990). Mechanisms of some chemical oscillators. *Journal of Physical Chemistry*, **94**, 4404–12.

Rabai, P., Orban, M., and Epstein, I.R. (1990). Design of pH-regulated oscillators. *Accounts of Chemical Research*, **23**, 258–63

Ruoff, P., Körös, E., and Varga, M. (1988). How bromate oscillators are controlled. *Accounts of Chemical Research*, **21**, 326–31.

Scott, S.K. (1987). Oscillations in simple models of chemical systems. *Accounts of Chemical Research*, **20**, 186–91.

Showalter, K. and Scott, S.K. (1992). Simple and complex propagating reaction-diffusion fronts. *Journal of Physical Chemistry*, **96**, 8702–11.

Swinney, H.L., Argoul, F., Arneodo, A., Richetti, P., and Roux, J.-C. (1987). Chemical chaos: from hints to confirmation. *Accounts of Chemical Research* **20**, 436–42.

Tyson, J.J. and Keener, J.P. (1988). Singular perturbation theory of travelling waves in excitable media (a review). *Physica*, **D32**, 327–61.

Winfree, A.T. (1974). Rotating chemical reactions. *Scientific American*, **230**, 82–95.

Books

Babloyantz, A. (1986). Molecules, dynamics and life: an introduction to self-organization of matter. Wiley, New York.

Field, R.J. and Burger, M. (eds) (1985). Oscillations and traveling waves in chemical systems. Wiley, New York.

Gray, P. and Scott, S.K. (1990). Chemical oscillations and instabilities: non-linear chemical kinetics. Oxford University Press.

Grindrod, P. (1991). Patterns and waves. Oxford University Press.

Glass, L. and Mackey, M.C. (1988). From clocks to chaos: the rhythms of life. Princeton University Press, New Jersey.

Hall, N. (ed.) (1992). The new Scientist guide to chaos. Penguin, London.

Murray, J.D. (1990). Mathematical biology. Springer, Berlin.

Nicolis, G. and Prigogine, I. (1989). Exploring complexity: an introduction. Freeman, New York.

Scott, S.K. (1991). Chemical chaos. Oxford University Press.

Stewart, I. (1989). Does God play dice? Backwell, Oxford.

Winfree, A.T. (1980). The geometry of biological time. Springer, Berlin.

Winfree, A.T. (1987). When time breaks down. Princeton University Press, New Jersey.

Index

acceleratory reaction 7
activation 40
activation energy 3
active media 41
AIDS 21
allosteric enzyme 40
attractor 35, 59, 81–3
   reconstructing 81
   strange 82
autocatalysis 7–9, 28
   cubic 8, 65
   mixed 9
   quadratic 8

basin of attraction 59
Belousov–Zhabotinsky (BZ)
   reaction 5–6, 9–10, 26–39
   in CSTR 87–90
   mechanism 27–9, 90
   recipe 26
   waves 41–6, 47–9
bifurcation 36, 54
   diagram 54
   Hopf 36
   saddle-node 54, 59
birhythmicity 62
bistability 53–4
brain cortex 50

calcium waves 50
cardiac arrhythmias 50
cell division 50
chain branching 10, 67–9
chaos 79–83
chemotaxis 49
clock reaction 14–18
CO oxidation 39, 75–83
   mechanism 78–9
concatenation 88
conservation condition 5
cool flame 39, 83–7
critical bromide ion concentration 28, 36
cross-shaped diagram 62–6
CSTR 53
curvature 46

deceleratory reaction 7
degenerate branching 86
degree of freedom 5, 59
detailed balance 11
dispersion relation 46

eikonal equation 47
elementary step 3
engine knock 87
excitability 37–9, 42
extent of reaction 7

Hodgkin–Huxley model 50
hydrogen-oxygen reaction 4–5, 10–11, 67–75
   closed vessel 67–72
   in CSTR 72–5
   first limit 70
   mechanism 67–9
   second limit 70–2
hysteresis 54

feedback 7–9
   thermal 9
fibrillation 50
filament 50
Fisher–Kolmogorov equation 21
FKN mechanism 27–9
flames 21–2
   cellular 25
flow
   branch 54
   diagram 56–8
   rate 54
   reactor 53
fractal 82
fronts 18–25
   nonplanar 22–5
   oscillatory 22
   planar 18–22
   speed 20–1

gas evolution oscillator 39
gasless pyrotechnics 23
glow
   steady 75, 77
   oscillatory 76, 77
glycolysis 40

heat balance equation 17
hydrocarbon oxidation 16, 83–7

independent concentration 5
induction period 14
inhibition 28, 40, 73
isola 64

Landolt reaction
   clock 14–16
   recipe 14

in CSTR 53–9
mechanism 11–12
oscillatory 63–4
law of mass action 3
limit cycle 34

mass balance equation 54–7
meander 48
Michaelis–Menten 40
mixed mode oscillation 80
mushroom 64

negative temperature coefficient 84
nerve transmission 50
net branching factor 10, 68
next-amplitude map 81
nonlinearity 6–7
nullcline 33

oregonator model 29–31, 43–6
   analysis of 31–3
   oscillations 33–7

pattern formation 50–2
period-doubling 80
phase plane 33
pH oscillator 64
pic d'arrêt
population models 21
predictability 83
pre-exponential factor 3

quasi-equilibrium 9

rabies 21
rate constant 3
rate determining step 10
rate law 4
   empirical 5
reaction
   acceleratory 7
   Belousov–Zhabotinsky (BZ) 5–6, 9–10, 26–39
   benzaldehyde oxidation 40
   Bray–Liebhafsky (BL) 39
   Briggs–Rauscher (BR) 39
   bromate-ferroin 21
   chlorite-iodide-malonic acid (CIMA) 39, 51

chlorite-thiosulfate 18
CO oxidation 39, 75–83
deceleratory 7
Dushman 11
hydrogen-oxygen 4–5, 10–11, 67–75
iodate-iodide-reductant 11–12
nitrate-ferroin 21
peroxidase-oxidase 40
Roebuck 11
recovery 46
refractory period 39
relaxation waveform 26
residence time 55

self-heating 9, 16
Semenov model 16–18
sickle cell anaemia 18
slime mould 49
steady state 53–4
   approximation 10
stoichiometric factor 4
superadiabatic temperature 23
supercatalysis 18

tangency 57–8
target patterns 43
thermal diagram 17
thermal explosion 16–18
thermal feedback 9
thermodynamic branch 53
thermokinetic mechanism 85–6
third body 71
threshold perturbation 37
trajectory 33
transition 58
Turing pattern 50–2
turning point 57

ventricular flutter 50

washout 64
wave 18–25, 41–50
   front 18–25
   kinematic 41
   phase 41
   pulse 41–2
   scroll 50
   speed 20–1, 42
   spiral 47–9
   target 43
   trigger 41